RIVERS AND WATER MANAGEMENT

Garrett Nagle

Dedication

To Angela, Rosie, Patrick and Bethany

Acknowledgements

The publishers would like to thank the following individuals, institutions and companies for permission to reproduce copyright illustrations in this book:

Hodder & Stoughton for extracts from *AS Geography: Concepts and Cases* by P Guinness and G Nagle (2000) used on pages 9 and 74; Kogan Page/Earthscan for an extract from *State of the World 2000* by L Brown (2000) used on page 133. All photographs, Garrett Nagle.

The publishers would also like to thank the following for permission to reproduce material in this book:

© *The Guardian* for an extract from *Saddam's troublesome marsh dwellers left high and dry by drainage* by R McCarthy (6 January 2003) used on page 49; © *The Guardian* for an extract from *Filthy water poisons the people of Dhaka's festering slums* by J Vidal (23 March 2002) used on page 127.

Every effort has been made to trace and acknowledge ownership of copyright. The publishers will be glad to make suitable arrangements with any copyright holders whom it has not been possible to contact.

Note about the Internet links in the book. The user should be aware that URLs or web addresses change regularly. Every effort has been made to ensure the accuracy of the URLs provided in this book on going to press. It is inevitable, however, that some will change. It is sometimes possible to find a relocated web page, by just typing in the address of the home page for a website in the URL window of your browser.

Orders: please contact Bookpoint Ltd, 130 Milton Park, Abingdon, Oxon OX14 4SB. Telephone: (44) 01235 827720. Fax: (44) 01235 400454. Lines are open from 9.00–6.00, Monday to Saturday, with a 24 hour message answering service. You can also order through our website www.hodderheadline.co.uk.

British Library Cataloguing in Publication Data
A catalogue record for this title is available from the British Library

ISBN 0 340 846 35 6

First Published 2003
Impression number 10 9 8 7 6 5 4 3 2 1
Year 2007 2006 2005 2004 2003 2002 2001

Cover photo: the Itaipú hydro-electric project on the Paraná River, Paraguay (by Michael Hill)
Produced by Gray Publishing, Tunbridge Wells, Kent
Printed in Great Britain for Hodder & Stoughton Educational, a division of Hodder Headline Plc, 338 Euston Road, London NW1 3BH by Bath Press Ltd.

Contents

1 The Hydrological Cycle

KEY DEFINITIONS

Hydrology The study of water.
Water cycle The movement of water between air, land and sea. It varies from place to place and over time.
Precipitation Includes all forms of rainfall, snow, frost, hail and dew. It is the conversion and transfer of moisture in the atmosphere to the land.
Interception The precipitation that is collected and stored by vegetation.
Overland runoff Water that flows over the land's surface.
Infiltration Water that seeps into the ground.
Evaporation Water from the ground or a lake that changes into a gas.
Transpiration Water loss from vegetation to the atmosphere.
Evapotranspiration The combined losses of transpiration and evaporation.
Hydrological cycle The movement of water between air, land and sea. It varies from place to place and over time.

1 Introduction

The study of water – hydrology – is an increasingly important one. Freshwater is a valuable resource and, despite there being plenty of it, it is not always available when or where it is needed. As the world's population grows and standards of living increase, the demand for water will continue to rise. To what extent the world's population can manage its water resources and water environments sustainably is a key issue.

In this chapter we consider the hydrological cycle at the global scale as well as some localised examples of human impacts on certain parts of the hydrological cycle. The hydrological cycle is, in theory, a closed system. However, some of the impacts on the cycle have a negative impact on its quality thereby limiting the capacity to recycle the water. As we will see throughout this book, management of all aspects of the hydrological cycle, and at all scales, is needed if we are to achieve sustainable development.

2 The Global Hydrological Cycle

The hydrological cycle refers to the movement of water between atmosphere, lithosphere and biosphere (Figure 1). At a global scale,

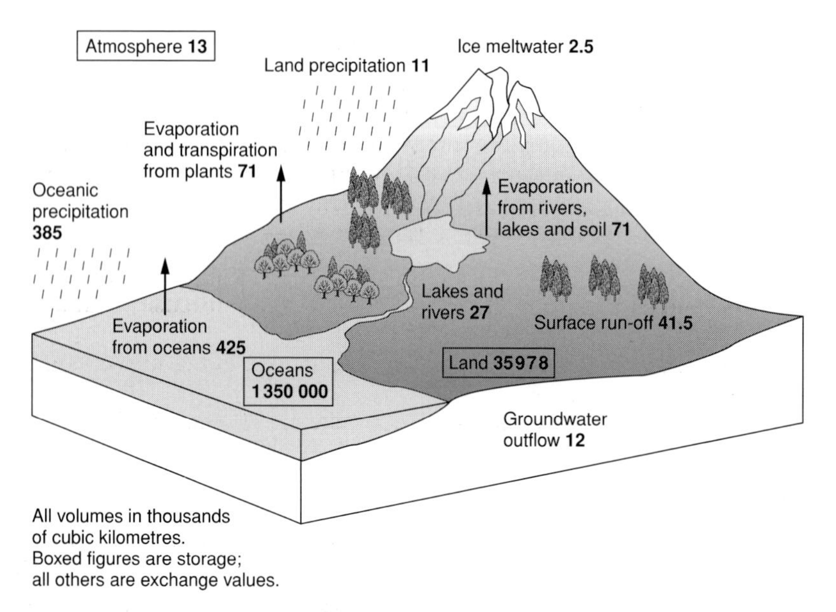

Figure 1 Hydrological cycle

it can be thought of as a closed system with no losses from the system. In contrast, at a local scale the cycle has a single input, precipitation (PPT), and two major losses (outputs), evapotranspiration (EVT) and runoff.

Water can be stored at a number of places within the cycle. These stores include vegetation, surface, soil moisture, groundwater and water channels. The global hydrological cycle also includes stores in the oceans and the atmosphere.

Human modifications are made at every scale. Good examples include large-scale changes of channel flow, irrigation and drainage, and abstraction of groundwater and surface water for domestic and industrial use.

a) Precipitation

Precipitation is considered in detail in a companion volume *Climate and Society*. Here it is important to mention the main characteristics that affect local hydrology. These are the amount of precipitation, the seasonality, intensity, type (snow, rain, etc.), geographic distribution and variability. For rain to occur, three factors must be satisfied:

- air is saturated, i.e. it has a relative humidity of 100%
- it contains particles of soot, dust, ash, ice, etc.

- its temperature is below dew point, i.e. the temperature at which the relative humidity is 100%, saturation is complete and clouds form.

Clouds are tiny droplets suspended in air, whereas rain droplets are much larger. Therefore cloud droplets must get much larger, although not necessarily by normal condensation processes. There are a number of theories to suggest how rain drops are formed.

There are three main types of rainfall:

- cyclonic – uplift of air within a low-pressure area (warm air rises over cold air): it normally brings low–moderate intensity rain and may last for a few days
- orographic – a deep layer of moist air is forced to rise over a range of hills or mountains
- convectional heating causes pockets of air to rise and cool.

b) Interception

Interception refers to water stored by vegetation. There are three main components:

- interception loss – water which is retained by plant surfaces and later evaporated away or absorbed by the plant
- throughfall – water which either falls through gaps in the vegetation or drops from leaves, twigs or stems
- stemflow – water which trickles along twigs and branches and finally down the trunk.

Interception loss from vegetation is usually greatest at the start of a storm following a dry period. This is due to:

- the interception capacity of the vegetation cover – when leaves and twigs are dry this is high, whereas as leaves become wetter the weight of water reduces surface tension and causes throughfall
- meteorological conditions especially windspeed may also decrease interception loss as intercepted rain is dislodged
- rainfall total is not important but the relative importance of interception losses will decrease as the amount increases
- the more frequent the storm events the less interception loss.

Interception loss varies with vegetation. For example, in German beech forests in summer it is up to 40% of rainfall, whereas in winter only about 20%. Interception is less from grasses than from deciduous woodland. Interception losses are greater from coniferous forests because:

- pine needles allow individual accumulation
- freer air circulation allows more evapotranspiration.

From agricultural crops, and from cereals in particular, interception increases with crop density.

Surface	Albedo (%)
Water (sun's angle over 40°)	2–4
Water (sun's angle less than 40°)	6–80
Fresh snow	75–90
Old snow	40–70
Dry sand	35–45
Dark, wet soil	5–15
Dry concrete	17–27
Black road surface	5–10
Grass	20–30
Deciduous forest	10–20
Coniferous forest	5–15
Crops	15–25
Tundra	15–20

Figure 2 Albedo rates

c) Evaporation

Evaporation is the process by which a liquid or a solid is changed into a gas. Its most important source is from oceans and seas. Evaporation increases under warm, dry windy conditions. When the air is warm the **saturation vapour pressure (*E*)** of water is high. By contrast, when the air is dry, the **actual vapour pressure (*e*)** of water in the air is low. Therefore the **saturation deficit (*E* – *e*)** is large. Increased saturation deficit leads to an increased evaporation rate.

Factors affecting evaporation include temperature, humidity and windspeed. Of these, temperature is the most important factor. Other factors include water quality, depth of water, size of water body, vegetation cover and colour of the surface (albedo or reflectivity of the surface, Figure 2).

d) Evapotranspiration

Transpiration is 'the process by which water vapour is transferred from vegetation to the atmosphere'. The combined effects of evaporation and transpiration are normally referred to as evapotranspiration (EVT). EVT represents the most important aspect of water loss, accounting for the removal of nearly 100% of the annual precipitation in arid areas and 75% in humid areas. Only over ice and snow fields, bare rock slopes, desert areas, water surfaces and bare soil will purely evaporative losses occur.

e) Potential evapotranspiration (P.EVT)

The distinction between actual EVT and P.EVT lies in the concept of moisture availability. Potential evapotranspiration is the water loss that would occur if there was an unlimited supply of water in the soil for use by the vegetation. Rates of potential evapotranspiration for the British Isles are shown in Figure 3(a) and actual evapotranspiration for England and Wales in Figure 3(b).

f) Infiltration

Infiltration is the process by which water soaks into or is absorbed by the soil. By contrast, percolation is the downward flow of water towards the water table. The two are closely related. The **infiltration**

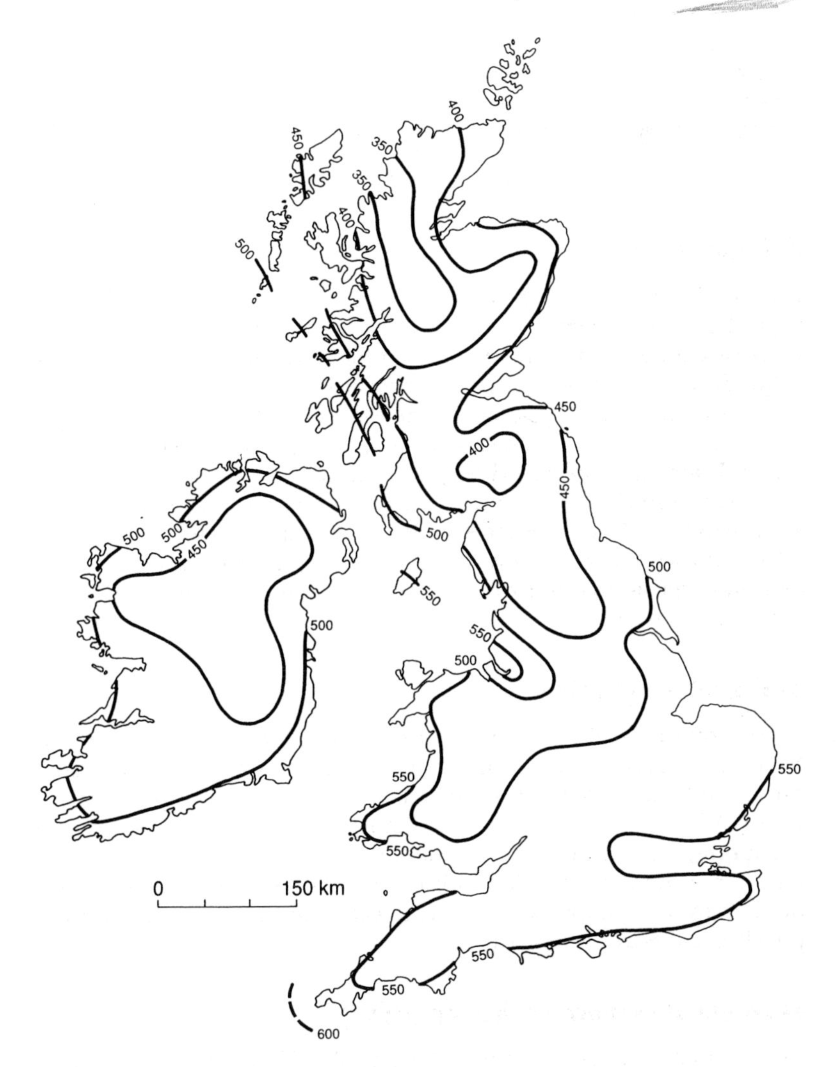

Figure 3a Potential evapotranspiration in the British Isles

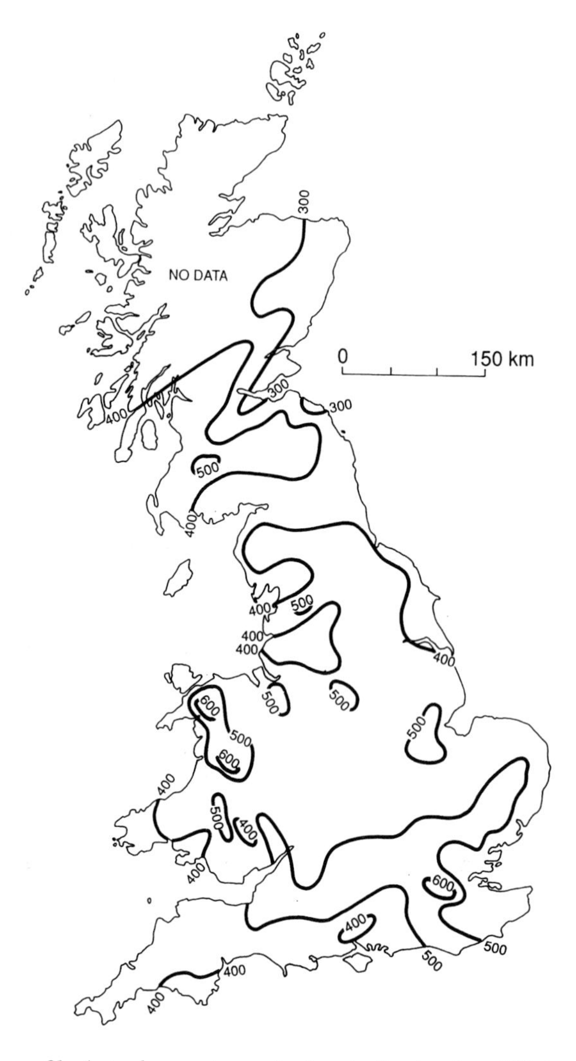

Figure 3b Actual evapotranspiration in England and Wales

capacity is the maximum rate at which rain can be absorbed by a soil in a given condition.

Infiltration capacity decreases with time through a period of rainfall until a more or less constant value is reached (Figure 4). Infiltration rates of 0–4 mm/hour are common on clays, whereas 3–12 mm/hour are common on sands. Vegetation also increases infiltration. On bare soils where rainsplash impact occurs, infiltration rates may reach 10 mm/hour. On similar soils covered by vegetation rates of between

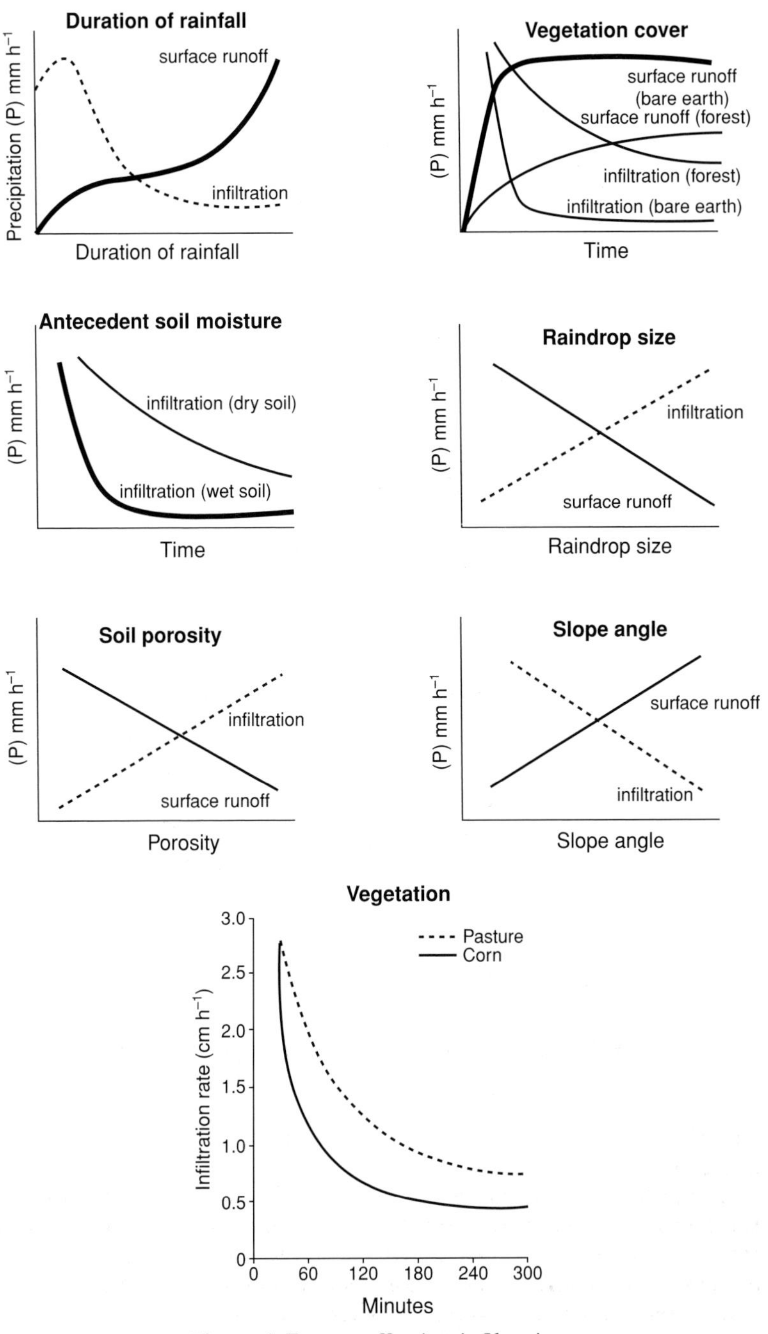

Figure 4 Factors affecting infiltration

50 and 100 mm/hour have been recorded. Infiltrated water is chemically rich as it picks up minerals and organic acids from vegetation and soil.

Infiltration is inversely related to overland runoff and is influenced by a variety of factors such as duration of rainfall, antecedent soil moisture (pre-existing levels of soil moisture), soil porosity, vegetation cover, raindrop size and slope angle (Figure 4).

g) Soil moisture

Soil moisture is the subsurface water in soil and subsurface layers above the water table. From here water may be:

- absorbed
- held
- transmitted downwards towards the water table, or
- transmitted upwards towards the soil surface and the atmosphere.

In coarser textured soils much of the water is held in fairly large pores at fairly low suctions, while very little is held in small pores. In the finer textured clay soils the range of pore sizes is much greater and, in particular, there is a higher proportion of small pores in which water is held at very high suctions.

Field capacity refers to the amount of water held in the soil after excess water drains away, i.e. saturation or near saturation. **Wilting point** refers to the range of moisture content in which permanent wilting of plants occurs.

There are important seasonal variations in soil moisture budgets (Figure 5):

- **Soil moisture deficit** is the degree to which soil moisture falls below field capacity. In Britain during late winter and early spring, soil moisture deficit is very low, due to high levels of precipitation and limited evapotranspiration.
- **Soil moisture recharge** occurs when precipitation exceeds potential EVT – there is some refilling of water in the dried-up pores of the soil.
- **Soil moisture surplus** is the period when soil is saturated and water cannot enter, and so flows over the surface.
- **Soil moisture utilisation** is the process by which water is drawn to the surface through capillary action.

h) Groundwater

Groundwater refers to subsurface water. The permanently saturated zone within solid rocks and sediments is known as the **phreatic zone**, and here nearly all the pore spaces are filled with water. The upper layer of this is known as the water table. The water table varies seasonally in Britain – it is higher in winter following increased levels of

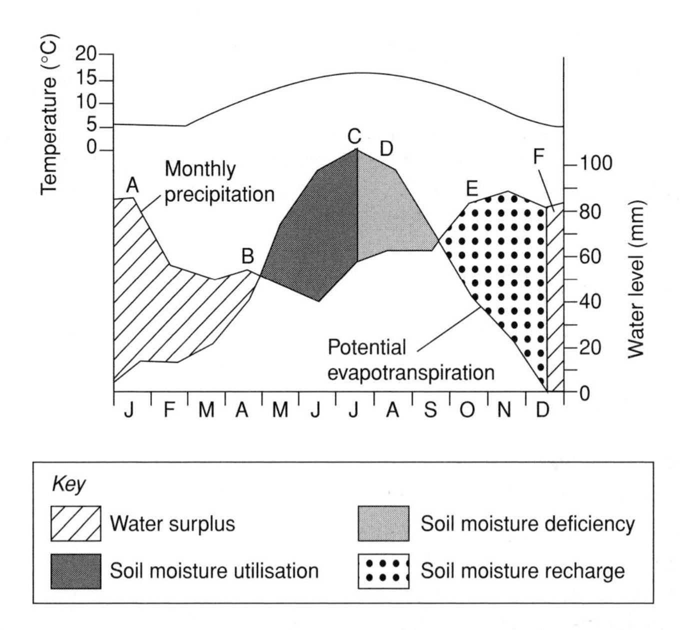

A Precipitation > potential evapotranspiration. Soil water store is full and there is a soil moisture surplus for plant use, runoff and groundwater recharge.

B Potential evapotranspiration > precipitation. Water store is being used up by plants or lost by evaporation (soil moisture utilisation).

C Soil moisture store is now used up. Any precipitation is likely to be absorbed by the soil rather than produce runoff. River levels will fall or dry up completely.

D There is a deficiency of soil water as the store is used up and potential evapotranspiration > precipitation. Plants must adapt to survive; crops must be irrigated.

E Precipitation > potential evapotranspiration. Soil water store will start to fill again (soil moisture recharge).

F Soil water store is full. Field capacity has been reached. Additional rainfall will percolate down to the water table and groundwater stores will be recharged.

Figure 5 Soil moisture status (reproduced by permission of Hodder Arnold)

precipitation. The zone that is seasonally wetted and seasonally dries out is known as the **aeration zone** or the **vadose zone**. Most groundwater is found within a few hundred metres of the surface but has been found at depths of up to 4 km beneath the surface (Figure 6).

Groundwater accounts for 96.5% of all freshwater on the earth. However, while some soil water may be recycled within a matter of days or weeks, groundwater may not be recycled for as long as 20,000 years. Hence, in some places, groundwater is considered a non-renewable resource.

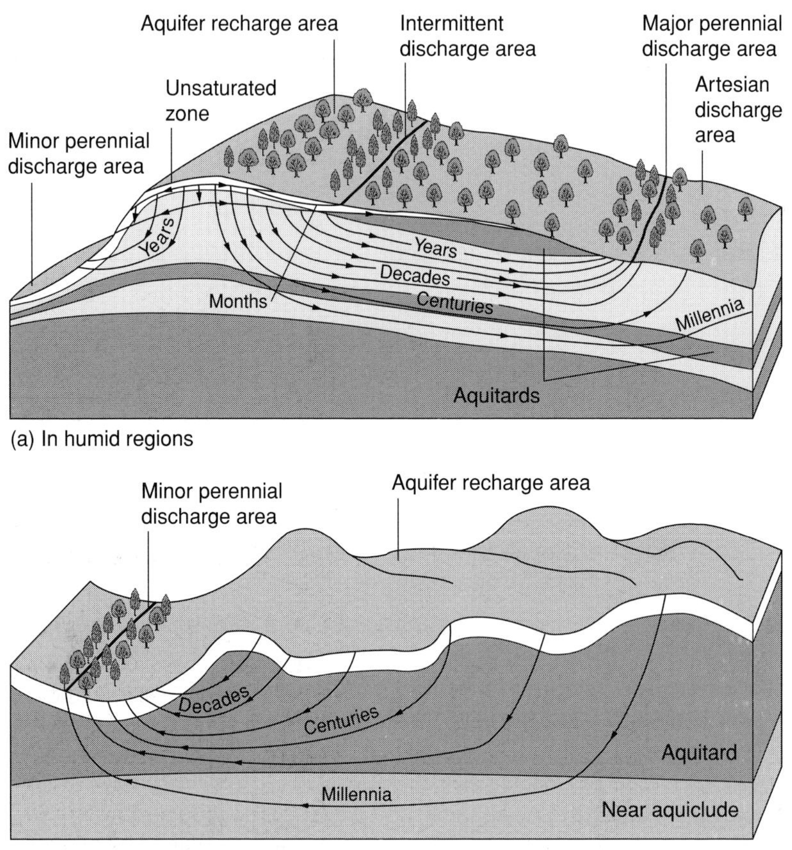

Figure 6 Groundwater

Aquifers (rocks which contain significant quantities of water) provide a great reservoir of water. Aquifers are permeable rocks such as sandstones and limestones. This water moves very slowly and acts as a natural regulator in the hydrological cycle by absorbing rainfall which otherwise would reach streams rapidly. In addition, aquifers maintain stream flow during long dry periods. A rock which will not hold water is known as an **aquiclude** or **aquifuge**. These are impermeable rocks that prevent large-scale storage and transmission of water, such as clay.

The groundwater balance is shown by the formula

$$\Delta S = Q_r - Q_d$$

where ΔS is the change in storage (+ or –), Q_r is recharge to groundwater and Q_d is discharge from groundwater.

Groundwater recharge occurs as a result of:

- infiltration of part of the total precipitation at the ground surface
- seepage through the banks and bed of surface water bodies such as rivers, lakes and oceans
- groundwater leakage and inflow from adjacent aquicludes and aquifers
- artificial recharge from irrigation, reservoirs, etc.

Losses of groundwater result from:

- evapotranspiration particularly in low-lying areas where the water table is close to the ground surface
- natural discharge by means of spring flow and seepage into surface water bodies
- groundwater leakage and outflow through aquicludes and into adjacent aquifers
- artificial abstraction, e.g. London Basin.

CASE STUDY: GROUNDWATER POLLUTION IN BANGLADESH

There has been an increase in the incidence of cancers in Bangladesh. It has been caused by naturally occurring arsenic in groundwater pumped up through the tube wells. Estimates by the World Health Organisation suggest that as many as 85 million of its 125 million population will be affected by arsenic-contaminated drinking water.

For 30 years, following the lead of UNICEF, Bangladesh has sunk millions of tube wells, providing a convenient supply of drinking water free from the bacterial contamination of surface water that was killing one-quarter of a million children a year. But the water from the wells was never tested for arsenic contamination, which occurs naturally in the groundwater. One in 10 people who drink the water containing arsenic will ultimately die of lung, bladder or skin cancer. Arsenic is tasteless, odourless, colourless and, taken in small doses in water, has no immediate ill-effects.

The first cases of arsenic-induced skin lesions were identified across the border in West Bengal, India, in 1983.

Arsenic poisoning is a slow disease. Skin cancer typically occurs 20 years after people start ingesting the poison. The real danger is internal cancers, especially of the bladder and lungs, which are usually fatal. Bangladeshi doctors have been warned to expect an epidemic of cancers by 2010. The victims will be people in their 30s and 40s who have been drinking the water all their lives – people in their most productive years.

The WHO says the recent discovery of arsenic-contaminated well water in Thailand, Taiwan, the United States, Argentina

Chile, Mexico, China, India and now Bangladesh shows it is a global problem. One solution to the problem is a concrete butt collecting water by pipe from gutters. Bangladesh's monsoon provides massive amounts rain that have not traditionally been collected for drinking. Other possible solutions include a simple filter using three ceramic pots suspended above each other, and a pump set up by a pond or river in which the pumped-up groundwater is filtered before emerging through a tap. But not one of the solutions being implemented experimentally is half as convenient as the tube wells it is designed to replace. Tube wells are easy to sink in the delta's soft alluvial soil, and for tens of millions of peasants the wells have revolutionised access to water.

Adapted from *The Independent*, 11 October 2000

3 The Influence of Human Activity

a) The human impact on precipitation

CASE STUDY: REASSESSING THE LYNMOUTH FLOODS, AUGUST 1952

A Government report released in 2002 suggests that the Lynmouth floods of August 1952 might not necessarily have been entirely natural. Weeks of heavy rain fell across Devon. On August some 250 mm of rain fell in just 12 hours. The water flowed down steep, but saturated, valleys in a raging torrent into the coastal town of Lynmouth killing 34 people and destroying many buildings.

New evidence suggests that there was more to it than just heavy rain. Declassified papers reveal that secret rainmaking experiments by the military had been taking place on Exmoor. Aircraft showered clouds with silver iodide to induce rainfall. Whether the experiment ran out of control, or whether the freak rains were going to fall in any case, is debatable.

Some efforts have been more blatant. During the Vietnam war, the Americans launched Project Popeye, a secret mission to seed the tops of monsoon clouds and trigger phenomenal downpours that would wash away the Ho Chi Minh Trail used for ferrying supplies. For 5 years Vietnam, Cambodia and Laos were sprayed during the monsoons, and military intelligence claimed that rainfall was increased by a third in some places.

Seeding can go wrong. In 1947, meteorologists tried to kill off a dying hurricane out at sea by seeding the clouds. The following day, the hurricane suddenly gathered strength, swung round and hit Savannah, Georgia, causing extensive damage.

There are a number of ways in which human activity effects precipitation. Cloud seeding has probably been one of the more successful. Rain requires either ice particles or large water droplets. Seeding introduces silver iodide, solid CO_2 (dry ice) or ammonium nitrate to attract water droplets. The results are unclear, partly it might be related to chance, some of it may be natural, some may be coincidental. In Australia and the United States seeding has increased precipitation by 10–30% on a small scale and on a short-term scale, but the increase in precipitation in one place might decrease precipitation elsewhere and it might also lead to an increase in hail. In urban and industrial areas precipitation is often increased by up to as much as 10% due to increased cloud frequency and amount, and because of the addition of pollutants, the heat island effect and turbulence.

b) The human impact on evaporation

The human impact on evaporation and evapotranspiration is relatively small in relation to the rest of the hydrological cycle but is nevertheless important. There are a number of reasons:

(i) Land-use changes – for example pine trees intercept more than deciduous trees. Tropical afforestation leads to increased transportation, however it is very costly.

(ii) Dams – there has been an increase in evaporation since the construction of large dams. For example Lake Nasser behind the Aswan Dam looses up to a third of the water due to evaporation. Water loss can be reduced by chemical sprays, by building sand fill dams and by covering the dams. Human activity also has an increasing impact on surface storage. There is increased surface storage due to the building of large-scale dams. These dams are being built with an increasing size and volume and number. This leads to a number of effects:

- increased storage of water
- decreased flood peaks (a decline of 71% in Cheviots)
- low flows in rivers, for example the River Hodder in Lancashire, the flow declined 10% in winter, but 62% in summer
- decreased sediment yields (clear water erosion)
- decreased losses due to evaporation and seepage leading to changes in temperature and salinity of the water
- increased flooding of the land
- triggering of earthquakes
- salinisation, for example in the Indus Valley, Pakistan, 1.9 million hectares are severely saline and up to 0.4 million hectares are lost per annum to salinity
- large dams can cause local changes in climate

In other areas there is a decline in the surface storage, for example in urban areas water is channelled away over impermeable surfaces in to drains and gutters very rapidly.

(iii) Urbanisation leads to a huge reduction in evapotranspiration due to the lack of vegetation. There may be a slight increase in evaporation due to higher temperatures and increased surface storage.

c) Human impact on interception

Interception is determined by vegetation, type and density. In farmland areas cereals intercept less than broad leaves. Row crops leave a lot of soil bare. Deforestation leads to a reduction in evapotranspiration, an increase in surface runoff, decline of surface storage and a decline in time lag. Afforestation has the opposite, although the evidence does not necessarily support it. For example in parts of the Severn catchment sediment loads increased four times after afforestation. This was due to a combination of an increase in overland runoff, little ground vegetation, young trees, access route for tractors, and fire breaks and wind breaks. All of these allowed a lot of bare ground. However, only 5 years later the amount of erosion went down.

d) The human impact on infiltration and soil water

Human activity has a major impact on infiltration and soil water. Land-use changes are important. Urbanisation creates an impermeable surface with compacted soil, which reduces infiltration. Infiltration is up to five times greater under forests compared with grassland. This is because the forest channels water down their roots and stems. With deforestation there is reduced interception, increased soil compaction and more overland flow.

Land-use practices are also important. Grazing leads to a decline in infiltration due to compaction and ponding of the soil. By contrast, ploughing increases infiltration because it loosens soils. Waterlogging and salinisation are common if there is poor drainage. When the water table is close to the surface evaporation of water leaves salts behind and may form an impermeable duricrust.

4 Changing Groundwater

Before irrigation development started in the 1930s, the Texas High Plains groundwater system was stable, with long-term recharge equal to long-term discharge. However, groundwater is now being used at a rapid rate to supply **centre-pivot irrigation schemes**. In less than 50 years, the water table level has declined by 30–50 m in a large area to the north of Lubbock, and the area irrigated by each well is contracting as well yields are falling.

By contrast, in some industrial areas, recent reductions in industrial activity have led to less groundwater being taken out of the ground. As a result, groundwater levels in such areas have begun to

CASE STUDY: CHANGING HYDROLOGY OF THE ARAL SEA

The Aral Sea began shrinking in the 1960s when Soviet irrigation schemes took water from the Syr Darya and the Amu Darya, and greatly reduced the amount of water reaching the Aral Sea. By 1994, the shorelines had fallen by 16 m (nearly 50 feet), the surface area had declined by 50%, the volume by 75% but salinity levels had increased by 300% (Figure 7).

There are many problems. Increased salinity levels killed off the fishing industry. Moreover, ports such as Muynak are now tens of kilometres from the shore. Salt from the dry seabed has reduced soil fertility and frequent dust storms are ruining the region's cotton production. Drinking water has been polluted by pesticides and fertilisers, and the air has been affected by dust and salt. There has been a noticeable rise in respiratory and stomach disorders, and the region has one of the highest infant mortality rates in the former Soviet Union.

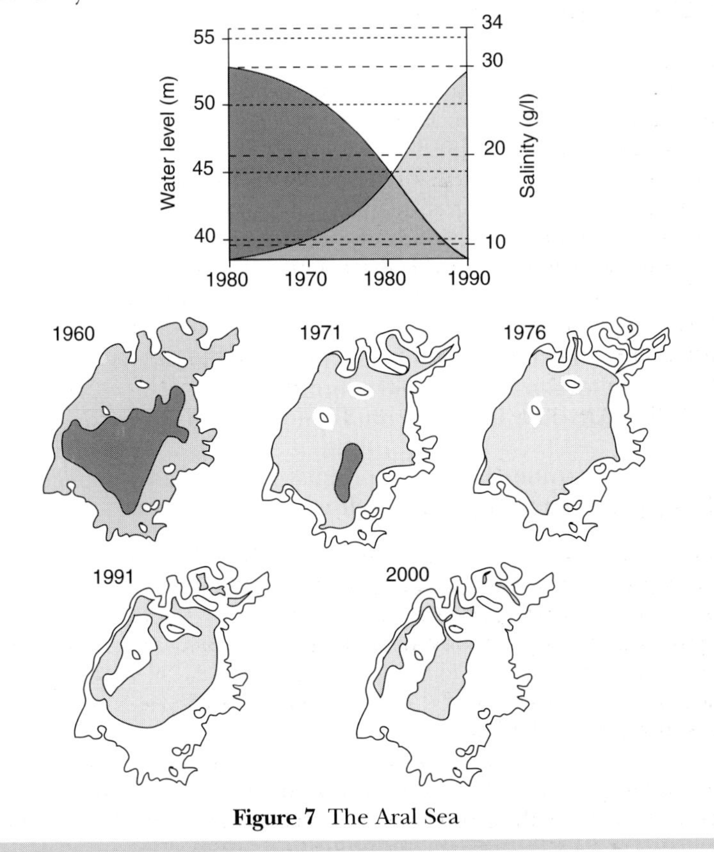

Figure 7 The Aral Sea

rise, adding to the problem caused by leakage from old, deteriorating pipe and sewer systems. This is happening in many British cities including London. Here, due to a 46% reduction in groundwater abstraction, the water table in the Chalk and Tertiary beds has risen by as much as 20 m. Such a rise has numerous implications including:

- an increase in spring and river flows
- surface water flooding
- pollution of surface waters and spread of underground pollution
- flooding of basements
- increased leakage into tunnels, e.g. tube lines
- reduction in stability of slopes and retaining walls
- swelling of clays as they absorb water
- chemical attack on building foundations.

There are a number of methods of recharging groundwater resources. Where the materials containing the aquifer are permeable water spreading is used. By contrast, in sediments with impermeable layers pumping water into deep pits or wells is used. This method is used extensively on the heavily settled coastal plain of Israel to replenish the groundwater reservoirs when surplus irrigation water is available, and to reduce the problems associated with salt-water intrusions from the Mediterranean.

Summary

- The hydrological cycle is a global closed system.
- A cycle is made up of inputs, processes and outputs.
- The main input is precipitation.
- The processes are varied and include interception, infiltration, overland runoff and base flow.
- Outputs include evaporation and transpiration (evapotranspiration) and at a local scale overland runoff.
- The outputs are linked to the inputs by evaporation and condensation.
- Human activities have had positive and negative impacts on all aspects of the hydrological cycle, and at a variety of scales.

Question on Hydrology

Define the following hydrological characteristics:

(i) interception
(ii) evaporation
(iii) infiltration.

Study Figure 4 which shows factors influencing infiltration and overland runoff. Write a paragraph on each of the factors, describing and explaining the effect it has on infiltration and overland runoff.

2 River Processes and Landforms

KEY DEFINITIONS

River regime Annual variation in the flow of a river.
Storm hydrograph Shows how a river changes over a short period, such as a day or a couple of days. Usually it is drawn to show how a river reacts to an individual storm. Each storm hydrograph has a series of parts.
Rising limb Shows how quickly the flood waters begin to rise.
Peak flow The maximum discharge of the river as a result of the storm.
Time lag The time between the height of the storm (not the start or the end) and the maximum flow in the river.
Recessional limb The speed with which the water level in the river declines after the peak.
Baseflow The normal level of the river, which is fed by groundwater.
Quickflow or **stormflow** The water that gets into the river as a result of overland runoff.
Abrasion The wearing away of the bed and bank by the load carried by a river.
Attrition The wearing away of the load carried by a river. It creates smaller, rounder particles.
Hydraulic action The force of air and water on the sides of rivers and in cracks.
Solution The removal of chemical ions, especially calcium.
Capacity The total load that a stream can carry.
Competence The size of the largest particle that a stream can carry.

1 Introduction

In this chapter we examine water flow in rivers at a number of scales – primarily at the annual scale (river regimes) and the short-term scale – flood hydrographs. We then consider the results of these flows, namely the processes of destruction, removal of debris and rebuilding. Increasingly, human activities are modifying these processes and the environments they create.

2 River Regimes

Stream flow occurs as a result of overland runoff, groundwater springs, from lakes and from meltwater in mountainous or subpolar

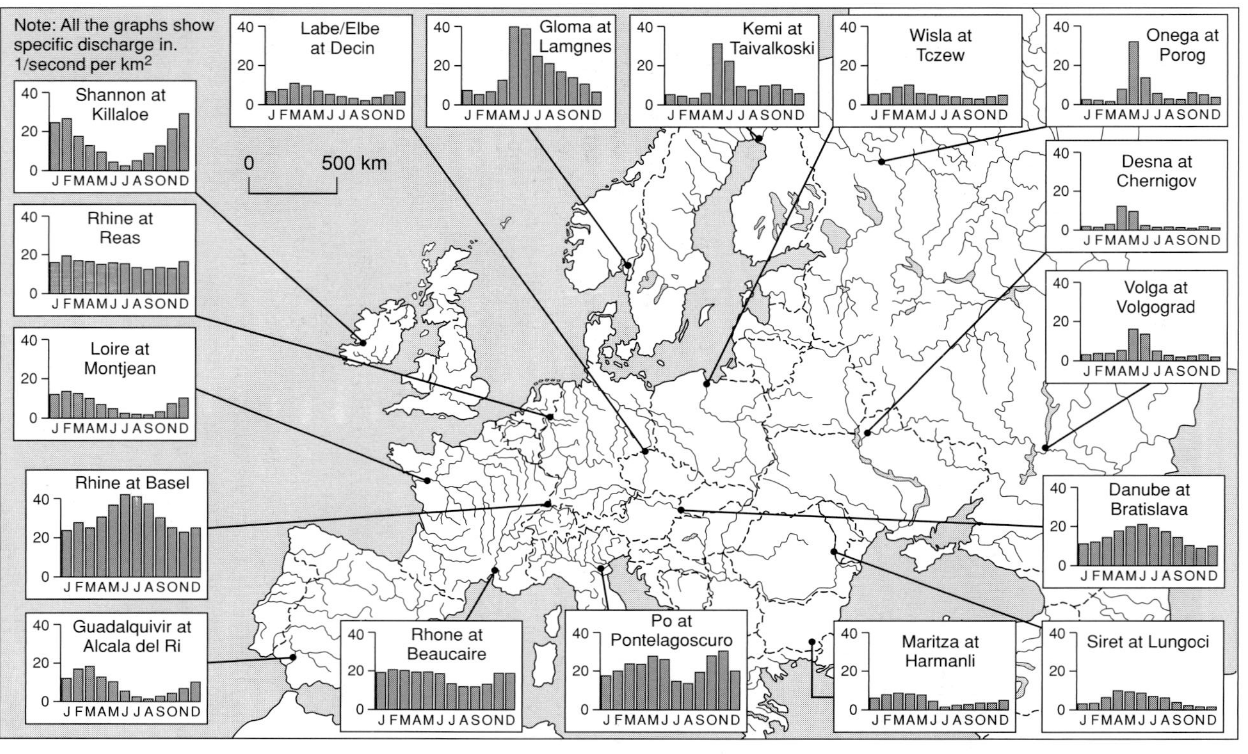

Figure 8 European regimes

environments. The character or **regime** of the resulting stream or river is influenced by several variable factors:

- the amount and nature of precipitation
- the local rocks, especially porosity and permeability
- the shape or morphology of the drainage basin, its area and slope
- the amount and type of vegetation cover
- the amount and type of soil cover.

On an annual basis the most important factor determining stream regime is climate. Figure 8 shows generalised regimes for Europe. Notice how the regime for the Shannon at Killaloe (Ireland) has a typical temperate regime, with a clear winter maximum. By contrast, Arctic areas such as the Gloma in Norway and the Kemi in Finland have a peak in spring, associated with snowmelt. Others such as the Po near Venice have two main maxima – autumn and winter rains (Mediterranean climate) and spring snowmelt from Alpine tributaries

3 Flood hydrographs

A flood hydrograph shows how the discharge of a river varies over a short time (Figure 9). Normally it refers to an individual storm or group of storms of not more than a few days in length. Before the storm starts the main supply of water to the stream is through groundwater flow or **baseflow**. This is the main supplier of water to rivers. During the storm some water infiltrates into the soil while some flows over the surface as overland flow or runoff. This reaches the river quickly as **quickflow**. This causes the rapid rise in the level of the river.

4 The Hydrology of Urbanisation

Geographers are increasingly aware of the effect of urbanisation on hydrology (Figure 10). In urban areas vegetated soils are replaced by impermeable surfaces. In some central areas, such as city centres, this replacement can be as high as 90%. By contrast, in areas of suburban detached housing, it can be as low as 5%. This change can lead to a number of effects:

- reduced water storage on the surface and in the soil
- increased percentage of runoff
- increases velocity of overland flow
- decreased evapotranspiration because urban surfaces are usually dry
- reduced percolation to groundwater because the surface is impermeable.

Second, there are major changes in the drainage density of urban areas. (Drainage density refers to the total length of stream channel

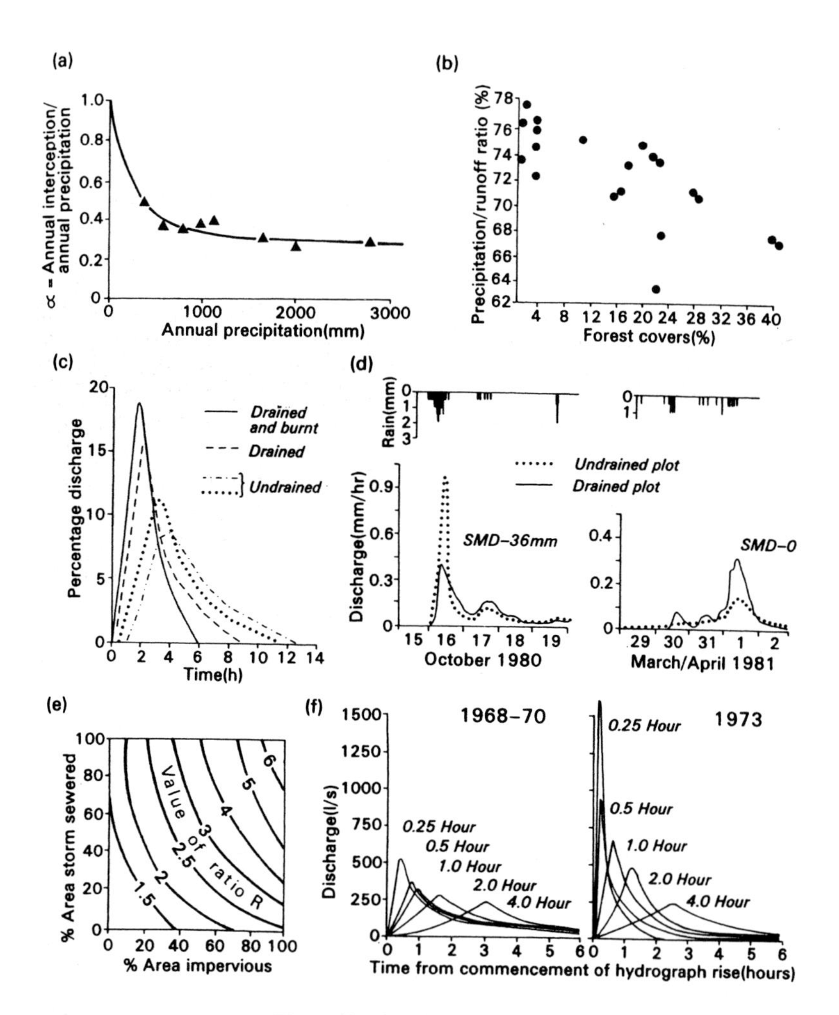

Figure 9 Flood hydrographs

per km².) The channel network is increased by storm water sewers, gutters, gullys and drains (Figure 11). Prior to urbanisation the stream channel network would have been much more limited. The increase in drainage density has a number of effects:

- it reduces the distance that overland flow has to travel before reaching the channel
- it increases the velocity of flow because sewers are smoother than natural channels
- it reduces storage in the channel system because sewers are designed to drain as completely as possible and as quickly as possible.

Urbanising influence	Potential hydrological response
Removal of trees and vegetation	Decreased evapotranspiration and interception; increased stream sedimentation
Initial construction of houses, streets and culverts	Decreased infiltration and lowered groundwater table; increased storm flows and decreased baseflows during dry periods
Complete development of residential, commercial and industrial areas	Decreased porosity, reducing time of runoff concentration, thereby increasing peak discharges and compressing the time distribution of the flow; greatly increased volume of runoff and flood damage potential
Construction of storm drains and channel improvements	Local relief from flooding; concentration of floodwaters may aggravate flood problems downstream

Figure 10 Potential hydrological effects of urbanisation

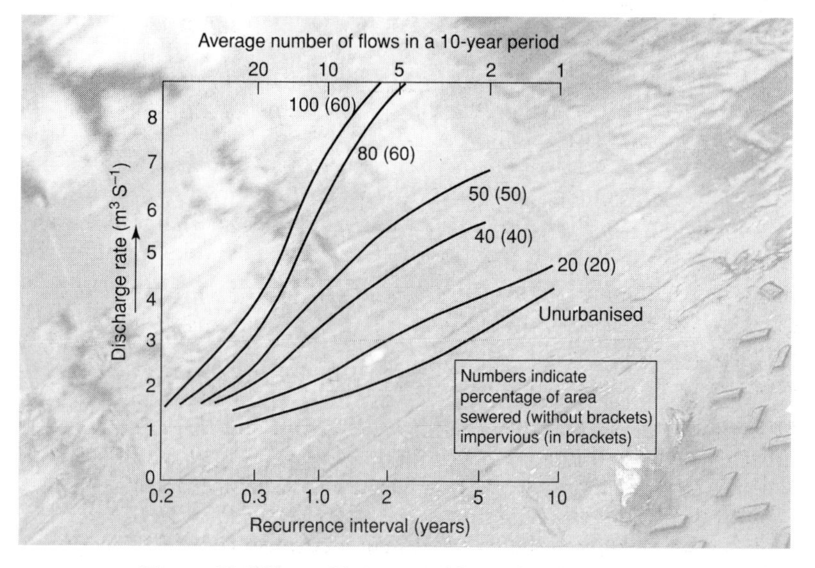

Figure 11 Effect of impermeable surfaces on runoff

In addition, there are rapid increases in rates of erosion during periods of construction. During the building of houses, roads and bridges vegetation is cleared. This exposes the soil to storms and allows increased amounts of overland flow. Heavy machinery disturbs and churns the soil, which increases its erodibility. However, some activities may bury the soil under concrete, tar or tiles. This effectively stops any further erosion of the soil.

The encroachment on the river channel by embankments reclamation and riverside roads has a number of consequences:

- channel width is reduced which leads to increased height of floods in the restricted channel
- bridges in the river can restrict the free discharge of floods and so increase flood levels upstream.

The combined effects of all these changes are that the flow regime, the flood hydrology, the sediment balance and the pollution load of streams are radically altered. Urbanisation is associated with increased peak flows. For example, a 243% increase resulted from the construction of Stevenage and an 85% rise followed the building of Skelmersdale. Likewise the peak of the hydrograph (resulting from 25 mm of rainfall) grew threefold and the lag time declined by 40% following the paving of an extra 6% of the basin of the Silk Stream in North London. Similarly, at Harlow in Essex the paving of 15% of the clay catchment increased total runoff by almost 60 mm and made it 130% of that of surrounding rural areas. The enlargement of river channels downstream of urban areas as a result of enlarged floods and reduced sediment discharges after the completion of building works have been demonstrated in the UK below Stevenage, Skelmersdale and Woodbury, Devon.

Urban storm water passes directly to open water courses via the storm sewers. By contrast, foul water passes to the sewage works via separate foul sewers or old-fashioned combined sewers. However, the storm water that washes off the roads and roofs of urban areas is not clean and unpolluted. Studies of urban streams show that during the start of urban runoff the quality of water can be worse than foul sewage. Such water may contain high levels of heavy metals, volatile solids and organic chemicals. These have been found in floods of the Silk Stream of London. Between 20% and 40% of storm-water sediments are organic in origin and most are biodegradable. By contrast, highway runoff has five to six times the concentration of heavy metals as roof runoff and the Silk Stream itself is severely polluted with faecal choliform bacteria. Annual runoff from 1 km of a single carriageway of the M1 included 1.5 tonnes of suspended sediment, 4 kg of lead, 126 kg of oil and 18 g of hazardous polynuclear aromatic hydrocarbons.

On all urban impermeable surfaces there is initial wetting of the surface, perhaps some absorption of water by the surfaces, certainly the filling of depressions in irregularities, and throughout and after virtually all rainstorms there must be evaporation from the surface into the atmosphere. Roofs are free from the infiltration of water but when the capacity of their gutters is exceeded in a severe storm there is overland flow to the ground. On the other hand, it is argued that after continued heavy rainfall there is no hydrological difference between tar and saturated soil. Beyond a certain threshold, which is hard to determine, the land use has little effect on flood magnitude.

5 Stream Flow

Stream flow and associated features of erosion are complex. The velocity and energy of a stream are controlled by:

- the gradient of channel bed
- the volume of water within channel, which is controlled largely by precipitation in drainage basin (e.g. 'bankfull' gives rapid flow whereas low levels give lower flows)
- the shape of the channel
- channel roughness, including friction

$$Q = (AR^{2/3} S^{1/2})/n$$

where Q = discharge, A = cross-sectional area, R = hydraulic radius, S = channel slope (as a fraction) and n = coefficient of bed roughness (the rougher the bed the higher the value).

If bed roughness increases velocity and discharge decrease, if the hydraulic radius and or slope/gradient increase the velocity and discharge will increase.

	Manning's n
Mountain stream, rocky bed	0.04–0.05
Alluvial channel (large dunes)	0.02–0.035
Alluvial channel (small ripples)	0.014–0.024

There are two main types of flow, laminar and turbulent. For laminar flow a smooth, straight channel with a low velocity is required. This allows water to flow in sheets or laminae parallel to the channel bed. It is rare in reality and most commonly occurs in the lower reaches. However, it is more common in groundwater and in glaciers when one layer of ice moves another.

Turbulent flow occurs where there are higher velocities and complex channel morphology such as a meandering channel with alternating pools and riffles. Turbulence causes marked variations in pressure within the water. As the turbulent water swirls (eddies) against the bed or bank of the river, air is trapped in pores, cracks and crevices and put momentarily under great pressure. As the eddy swirls away, pressure is released; the air expands suddenly, creating a small explosion that weakens the bed or bank material. Thus, turbulence is associated with hydraulic action (cavitation).

Vertical turbulence creates hollows in the channel bed. Hollows may trap pebbles, which are then swirled by eddying, grinding at the bed. This is a form of vertical corrosion or abrasion and, given time, may create pot-holes. Cavitation and vertical abrasion may help to deepen the channel, allowing the river to downcut its valley. If the downcutting is dominant over the other forms of erosion (i.e. vertical erosion exceeds lateral erosion) then a gulley or gorge will develop.

6 Erosion and Transport

There are a number of factors affecting rates of erosion. These include:

- **load** – the heavier and sharper the load the greater the potential for erosion
- **velocity** – the greater the velocity the greater the potential for erosion
- **gradient** – increased gradient increases the rate of erosion
- **geology** – soft, unconsolidated rocks such as sand and gravel are easily eroded
- **pH** – rates of solution are increased when the water is more acidic
- **human impact** – deforestation, dams and bridges interfere with the natural flow of a river and frequently end up increasing the rate of erosion.

Erosion by the river will provide loose material. This eroded material (plus other weathered material that has moved downslope from the upper valley sides) is carried by the river as its load. The load is transported downstream in a number of ways:

- the smallest particles (silts and clays) are carried in suspension as the **suspended load**
- larger particles (sands, gravels, very small stones) are transported in a series of 'hops' as the **saltated load**
- pebbles are shunted along the bed as the **bed or tracted load**
- in areas of calcareous rock, material is carried in solution as the dissolved load.

a) Global sediment yield

It is possible to convert a value of mean annual sediment and solute load to an estimate of the rate of land surfaces lowering by fluvial denudation. This gives a combined sediment and solute load of 250 tonnes/km^2/year, i.e. an annual rate of lowering of the order of 0.1 mm/year. There is a great deal of variation in sediment yields. These range from 10 tonnes/km^2/year in such areas as northern Europe and parts of Australia to in excess of 10,000 tonnes/km^2/year in certain areas where conditions are especially conducive to high rates of erosion (Figure 12). These include Taiwan, South Island New

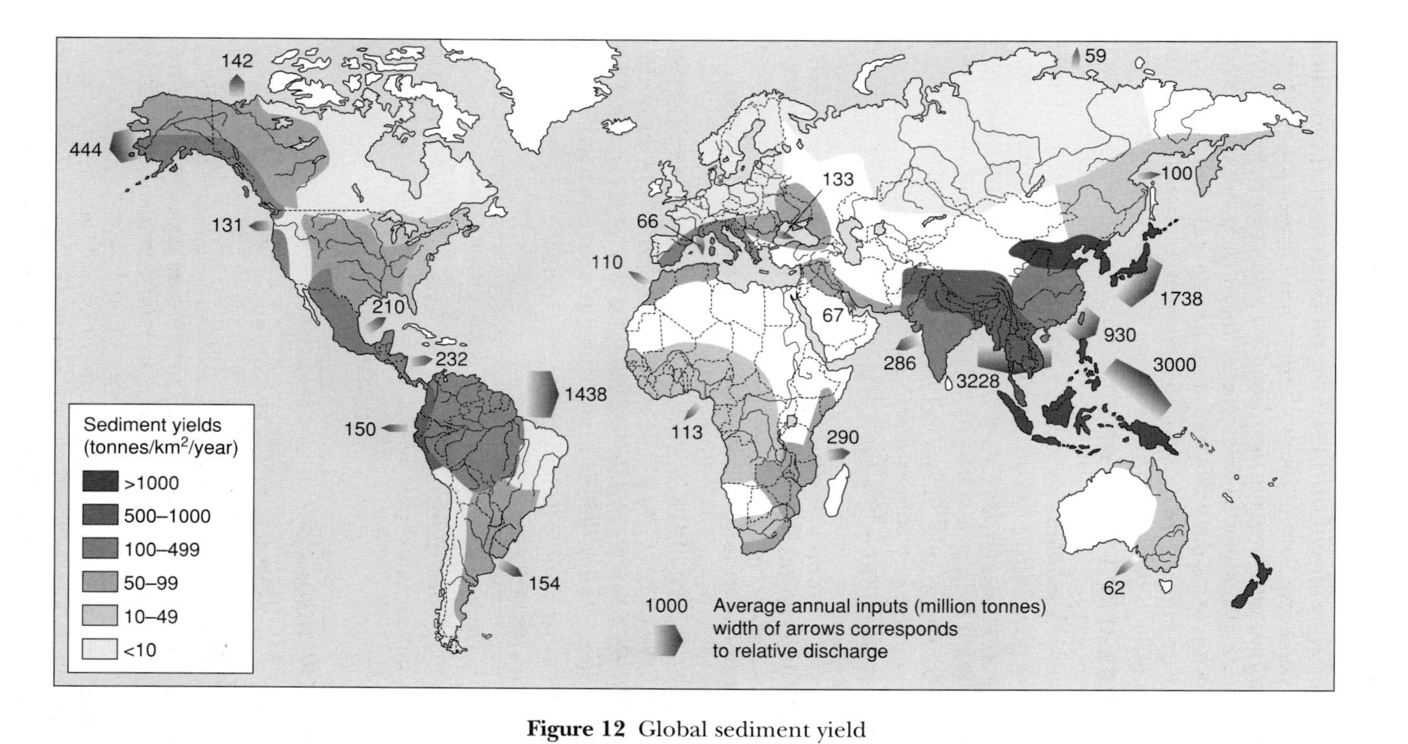

Figure 12 Global sediment yield

Zealand and the Middle Yellow River Basin in China. In the two former cases steep slopes, high rainfall and tectonic instability are major influences whilst in the latter case the deep loess deposits and the almost complete lack of natural vegetation cover are important. Rates of land surface lowering associated with this component of fluvial transport can be seen as varying over more than three orders of magnitude from less than 0.004 mm/year to in excess of 4 mm/year. The broad pattern of global suspended sediment is shown in the diagram and it reflects the influence of a wide range of factors including climate, relief, geology, vegetation cover and land use.

A survey of over 1500 recording stations has resulted in a map of global suspended sediment yield. Very high levels of sediment were found in Mediterranean areas, south-west USA and parts of East Africa. This were largely due to semi-arid climatic conditions, with irregular rainfall falling on a partial vegetation cover. In contrast, the high sediment levels for the Pacific Rim are the result of:

- high relief
- tectonic activity
- high rainfall.

In areas such as central Eurasia and North America, the combination of low relief and resistant rocks reduced sediment levels. In tropical Africa the rates were reduced further still by the dense covering of vegetation which intercepted most of the rainfall.

Large amounts of sediment are transported in rivers in south-east Asia and Oceania (Figure 13). These areas account for just 15% of the world's land surface but about 70% of the sediment carried by rivers. The highest rates for suspended sediment came from the Huangfuchuan River in

Country	River (km^2)	Drainage area (t/km^2/year)	Mean annual suspended sediment yield
China	Huangfuchaun	3199	53,500
China	Dali	96	25,600
Taiwan	Tsengwen	1000	28,000
Kenya	Perkerra	1310	19,520
Java	Cilutung	600	12,000
Java	Cikeruh	250	11,200
New Guinea	Aure	4360	11,126
New Zealand	Waiapu	1378	19,970
New Zealand	Waingaromia	175	17,340
New Zealand	Hikuwai	307	13,890

Figure 13 Maximum values of mean annual specific suspended sediments yield for world rivers

China, which had an average sediment yield of 53,000 tonnes/km/year. The reasons for such a huge load are varied are include:

- erodable sediments (such as loess, glacial deposits and volcanic ash)
- high relief
- tectonic activity
- erosive storms (tropical cyclones)
- limited vegetation cover
- intense human activity, removing vegetation and exposing ground surfaces.

Large amounts of sediment are also carried by mountain streams. In North America, for example, the loads of the Sustitna, Stekine and Yukon together carry more sediment than the mighty Mississippi. Similarly, in Europe, the little-known Semani River of Albania carries twice as much sediment as the Garonne, Loire, Seine, Rhine, Weser, Elbe and Oder combined. It is estimated that of the total sediment carried from land to sea each year – some 2.7–4.6 km^3 – more than 1.1 km^3 comes from the Himalayas.

b) Rates of erosion in Britain

There have been very few attempts to measure dissolved, suspended and bed loads at the same time (Figure 14). The results are mixed. Studies from the Pennines suggest that bed load is relatively unimportant, although dissolved load and suspended load varied in importance. By contrast, in the headwaters of the Upper Wye, bed load was seen to be more important than suspended load, although dwarfed by dissolved load.

Riverbank erosion is related to river discharge and width. Rates of erosion may be high enough to cause concern and practical problems for riparian communities (Figure 15). Riverbank erosion is an important source of suspended sediment in a river, and is an important part in the development of floodplains (caused by the migration of meanders).

Information on the dissolved loads of rivers at the global scale is less complete than that available for suspended sediment. Existing knowledge suggests loads ranging from less than 1 tonne/km^2/year to approximately 500 tonnes/km^2/year. These values are somewhat lower than those associated with suspended sediment transport, and if total material transport from the land surface of the Earth to the oceans is considered, the value for suspended sediment, which is in the order of 15×10^9 tonnes per year, exceeds that for dissolved material (4×10^9 tonnes per year) by nearly four times. As a global generalisation the relative efficiency of mechanical and chemical fluvial denudation could therefore be seen as being in the ratio of 4 to 1.

Location and load	Quantity (t/km^2/year)	Percentage
North Pennine Basins		
Carl Beck		
Dissolved	110.28	81.5
Suspended	24.77	18.3
Bed	0.33	0.2
Total	135.38	100.0
Great Eggleshape Beck		
Dissolved	0.0	0.0
Suspended	12.07	95.6
Bed	0.55	4.4
Total	12.62	100.0
Afon Cyff, Upper Wye, Mid-Wales		
Dissolved	37.6	77.7
Suspended	2.8	5.8
Bed	8.0	16.5
Total	48.4	100.0
Shropshire catchments		
Preston Montfort		
Dissolved	67–100	–
Suspended	32.8	–
Bed	1.05	–
Total	100.85–123.85	–
Farlow		
Dissolved	113.02	50.219
Suspended	112.03	49.780
Bed	0.0002	0.001
Total	225.052	100.00

Figure 14 Measurements of sediment load

River	Dates of survey	Rate (m/year)
Cound, Shropshire	1972–4	0.64
Bollin, Cheshire	1872–1935	0.16
Exe, Devon	1840–1975	2.58
Creedy, Devon	1840–1975	0.52
Culm, Devon	1840–1975	0.51
Axe, Devon	1840–1975	1.0
Yatty, Devon	1840–1975	1.38
Coly, Devon	1840–1975	0.48
Hookamoor, Devon	1840–1975	0.19
Severn	1948–75	0.2–0.7
Rheidol	1951–71	1.75
Tywi	1905–71	2.65

Figure 15 Rates of riverbank erosion

7 Features of Erosion

a) Gorges and waterfalls

Gorge development is common, for example, where the local rocks are very resistant to weathering but more susceptible to the more powerful river erosion. Similarly, in arid areas where the water necessary for weathering is scarce, gorges are formed by episodes of fluvial erosion. A rapid acceleration in downcutting is also associated when a river is rejuvenated, again creating a gorge-like landscape. Gorges may also be formed as a result of:

- antecedent drainage (Rhine Gorge)
- glacial overflow channelling (Newtondale)
- collapse of underground caverns in carboniferous limestone areas (River Axe at Wookey Hole)
- surface runoff over limestone during a periglacial period (Cheddar Gorge)
- retreat of waterfalls (Niagara Falls).

Plunge flow occurs where the river spills over a sudden change in gradient, undercutting rocks by hydraulic impact and abrasion, thereby creating a waterfall. There are many reasons for this sudden change in gradient along the river:

- a band of resistant strata such as the resistant limestones at Niagara Falls
- a plateau edge such as Livingstone Falls
- a fault scarp such as at Gordale
- a hanging valley such as at Glencoyne, Cumbria
- coastal cliffs, such as Kimmeridge Bay, Dorset.

The undercutting at the base of the waterfall creates a precarious overhang, which will ultimately collapse. Thus, a waterfall may appear to migrate upstream, leaving a gorge of recession downstream. The Niagara Gorge is 11 km long due to the retreat of Niagara Falls (Figure 16).

CASE STUDY: WATERFALLS

Niagara Falls

Most of the world's great waterfalls are the result of the undercutting of resistant cap rocks, and the retreat or recession that follows thereafter. The following case studies illustrate this clearly, although the third example, from Iceland, illustrates a different mode of formation.

The Niagara River flows for about 50 km between Lake Erie and Lake Ontario (Figure 16). In that distance it falls just 108 m giving an average gradient of 1:500. However, most of the descent occurs in the 1.5 km above the Niagara Falls (13 m) and at the Falls themselves (55 m). The Niagara River flows in a 2 km wide channel just 1 km above the Falls, and then into a narrow 400 m wide gorge, 75 m deep and 11 km long. Within the gorge the river falls a further 30 m.

The course of the Niagara River was established about 12,000 years ago when water from Lake Erie began to spill northwards into Lake Ontario. In doing so, it passed over the highly resistant dolomitic (limestone) escarpment. Over the last 12,000 years the Falls have retreated 11 km, giving an average rate of retreat of about 1 m/year. Water velocity accelerates over the Falls, and decreases at the base of the Falls. Hydraulic action and abrasion have caused the development of a large plunge pool at the base of the Falls, while the fine spray and eddies in the river help to remove some of the softer rock underneath the resistant dolomite. As the softer rocks are removed the dolomite is left unsupported and the weight of the water causes the dolomite to collapse. Hence the waterfall retreats forming a gorge of recession.

In the 19th century rates of recession were recorded at 1.2 m/year. However, now that the amount of water flowing over the

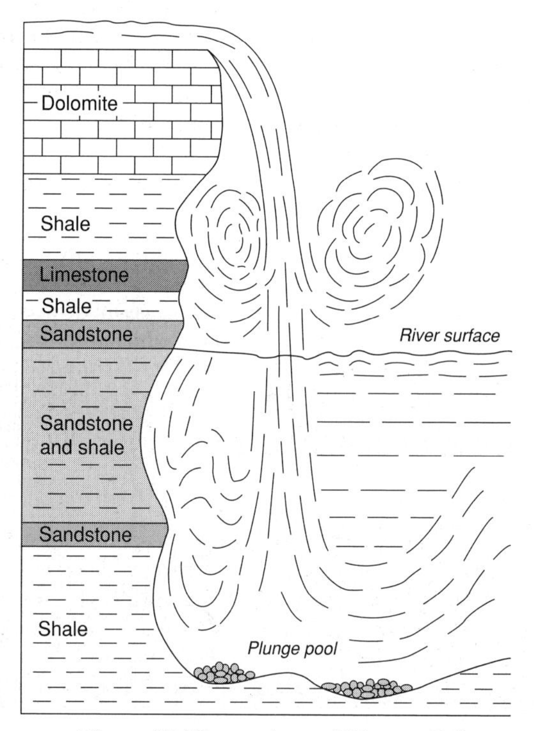

Figure 16 The geology of Niagara Falls

Falls is controlled (due to the construction of hydroelectric power stations) rates of recession have been reduced. In addition, engineering works in the 1960s reinforced parts of the dolomite that were believed to be at risk of collapse. The Falls remains an important tourist attraction, and local residents and business personnel did not want to loose their prized asset!

Victoria Falls

Victoria Falls, on the border of Zambia and Zimbabwe, is one of the world's most spectacular waterfalls (Figure 17). Like Niagara, it has a resistant cap rock, but, unlike Niagara, one of the dominant forces in the creation and development of Victoria Falls is plate tectonics. The Falls are nearly 2 km wide, up to 108 m deep, and during the rainy season over 5 million m^3/min pass over the Falls. During the dry season it is much less, but the annual average is still an impressive 550,000 m^3/min. When the river passes over the Falls its channel narrows from over 1.7 km to just a few metres. This causes an increase in the river's velocity which causes much erosion and scouring to occur.

The evolution of the Falls is complex. About 150 million years ago molten rock formed fine-grained resistant basalt, about 300 m thick, in the Victoria Falls area. As the lava cooled, it shrank, causing cracks or fissures to appear in the rock. These fissures were later widened by weathering, and were infilled by deposits of soft clay and lime. Over time these deposits solidified to form limestone. Continued tectonic processes caused large east–west fissures to widen, allowing the limestone in the fissures to be eroded.

Figure 17 Victoria Falls

The Zambezi's course changed over time. Uplift caused the ancient course of the river to be blocked, and diverting the river east into its present course. The river was able to erode the soft sandy deposits on the surface, but was unable to erode the harder basalt below. At the time the edge of the basalt plain was some 100 km from the present Victoria Falls. As the waterfall retreated it occupied an east–west fissure filled with relatively soft limestone. This it was able to erode easily and cut back to within 8 km of the present waterfall. Then it occupied a north–south fissure, until it cut back as far as the next east–west fissure. The present Victoria Falls is believed to be similar to the previous falls, each of which has occupied different fissures within the resistant basalt.

Gullfoss, Iceland: an example of cap rock, fault-controlled development and glacier bursts

The waterfall at Gullfoss in south-west Iceland is an excellent example of a waterfall than illustrates the interplay of rock type, faulting, sea-level change and glacial activity (Figure 18). As with Victoria Falls, resistant basalt forms the cap rock, while a mixture of conglomerates, moraines, sandstones and siltstones form the softer underlying rocks. A combination of high discharges and high velocities leads to erosion of the softer rock, leaving the harder basalt to protrude as a ledge. As the softer rocks are removed, the support for the basalt disappears and the waterfall retreats. However, there is much more than just this.

The water flowing over Gullfoss comes from the Langjokull (literally Long Glacier), and is therefore very seasonal in nature.

Figure 18 Gullfoss

Maximum velocities are generally in late spring and throughout summer, when melting of the glacier is at a maximum. Occasionally there are glacier bursts, or jokulhlaups. When this happens the discharge of the river increases from 130 cumecs (cubic metres per second) (the normal summer discharge) to over 2000 cumecs. Such a force has tremendous potential for erosion and destruction.

At Gullfoss, the river suddenly changes direction. This is because it follows a fault line. The fault line represents a line of weakness in the rock, and is easier to erode, as it is already fractured and cracked. Evidence of sea-level change is provided by the old terraces. These represent the old floodplains of the Hvita (literally 'White River'). When land rose at the end of the last glacial period the river had to readjust itself to a lower base level, and so cut through the former floodplain. If the level of the land relative to the sea changes again, then another terrace is formed.

Waterfalls and sea-level change

At Seljalandfoss in southern Iceland, the waterfall that can be seen is one that falls over an abandoned sea cliff (Figure 19). As the level of the land rose relative to the level of the sea, the cliffs were cut off from the sea, and are now located some 3–4 km from the sea, They are separated from the sea by an extensive sandur or outwash plain. There is no gorge of recession, just a relatively small river plunging 65 m over the former marine cliffs. Interestingly, farms are now located on some of the former islands (now exposed as small hills on the outwash plain). A scree slope is developing at the base of the cliff, the result of weathering, mass movement, and erosion during the spring and summer flow of water at Seljalandfoss.

Figure 19 Seljalandfoss

b) Helicoidal flow

Horizontal turbulence often takes the form of **helicoidal flow**, a 'corkscrewing' motion. This is associated with the presence of alternating pools and riffles in the channel bed, and where the river is carrying large amounts of material. The erosion and deposition by helicoidal flow creates meanders.

Traditionally, the study of meanders focused upon their shape and their role in the evolution of river valley. The processes involved received little attention. Initial process measurements focused on the rates of development as a surrogate for the processes, but increasing concern for mechanics and models has now generated a greatly improved understanding of the hydraulics of meanders, even if the factors ultimately responsible for them remain unclear.

Meanders occur as the top current impinges against the channel bank, undercutting it and causing bank caving. This leaves river cliffs at the channel edge. As the water heaps up against this bank area, a bottom or subsidiary current develops carrying energy and material to the opposite bank but slightly downstream. Thus, a corkscrew motion is established. Immediately opposite the area of bank caving there will be an area of slow water where deposition occurs, forming a slip-off slope. Hence, as one bank area is being eroded, the opposite bank area is being built-up – thus a meander is allowed to develop, with a characteristic asymmetric cross-section.

Once established, a meandering pattern is self-perpetuating. As a result, the meanders become accentuated, i.e. the amplitude of the bends are increased. Sometimes to such an extent that the meander swings round almost to meet itself, leaving only a 'swan neck' of land between. The flow round this much accentuated meander becomes increasingly tortuous and only a slight increase in volume or velocity will cause the water to flood the swan's neck, eroding a new, straighter channel, and abandoning the meander as an ox-bow lake (cutoff). Eventually, this ox-bow lake will silt up and be colonised by marshland vegetation.

As well as the change in amplitude, there is a change over time, in the wavelength of the meander. This increase in wavelength will cause: (a) the trimming of the intervening spurs and bluffs lying alongside the channel; (b) the erosion and reworking of the slip-off slope and other deposited material; and (c) the apparent shift or migration of the meander downstream.

These changes in the dimensions of the meanders allow the river to cut a trough-like area, with the meanders swinging from side to side, filling the floodplain. Gradually this trough will be enlarged by meander development, while the bluffs at the sides of the floodplain will be degraded by weathering and other surface processes of erosion. Thus, at the end of a very long time, the river will meander freely over the extensive plain.

8 Deposition

There are a number of causes of deposition such as:

- a shallowing of gradient which decreases velocity and energy
- a decrease in the volume of water in the channel
- an increase in the friction between water and channel.

a) Alluvial fans and cones

Many types of deposition are found along the course of a river. In general, as a river carries its load, the load is eroded by attrition. This results in the load getting rounder and smaller downstream (Figure 20). **Piedmont alluvial fans and cones** are found in semi-arid areas where swiftly flowing mountain streams enter a main valley or plain at the foot of the mountains. There is a sudden decrease in velocity causing deposition. Fine material is spread out as an alluvial fan with a nearly

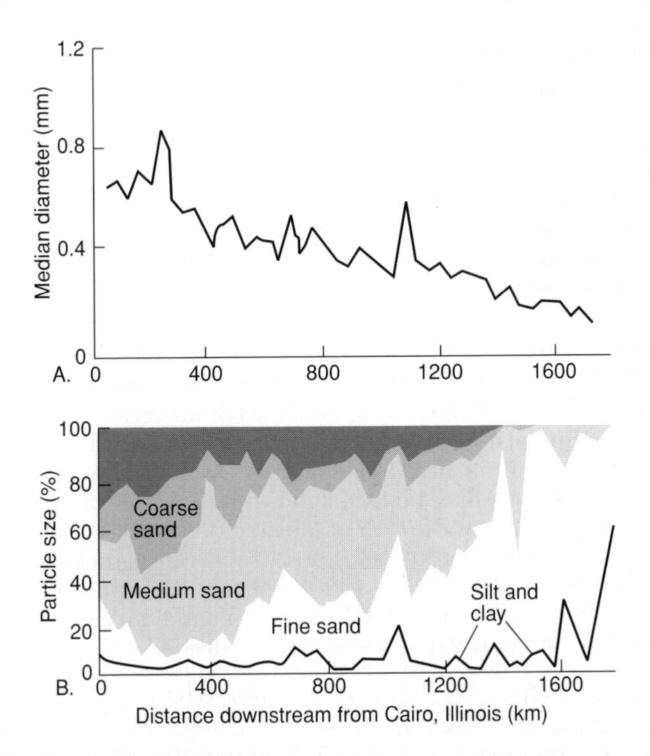

Figure 20 Changes in sediment size along the Mississippi River downstream from Cairo, Illinois. Over the lower 1600 km of the river, median diameter decreases from 0.7 mm to 0.2 mm. At Cairo, about 40% of the stream's load is gravel, 50% is sand, and 10% is silt and clay. At a point 1600 km downstream, the sediment is almost entirely finer than medium-grained sand

flat surface, under 1° angle. By contrast, coarse material forms a relatively small, steep-sided alluvial cone, with a slope angle of up to 15°. They are also common therefore in glaciated areas at the edges of major troughs, particularly at the base of hanging valleys. However, they are much smaller than their semi-arid counterparts.

b) Riffles

Riffles are small, ridges of material deposited where the river velocity is reduced midstream, in between pools (the deep parts of a meander). If many such ridges are deposited, the river is said to be 'braided'. A braided river channel consists of a number of interconnected shallow channels separated by alluvial and shingle bars (islands). These may be exposed during low flow conditions. They are formed in rivers that are heavily laden with sediment and have a pronounced seasonal flow. There are excellent examples on the Eyra Fjordur in northern Iceland.

c) Levees and floodplains

Levees and floodplain deposits are formed when a river bursts its banks over a long period of time. Water quickly looses velocity, leading to the rapid deposition of coarse material (heavy and difficult to move a great distance) near the channel edge. These coarse deposits build up to form embankments called **levees**. The finer material is carried further away to be dropped on the **floodplain**, sometimes creating **backswamps**.

d) Deltas

Deltas are river sediments deposited when a river enters a standing body of water such as a lake, a lagoon, a sea or an ocean. They are the result of the interaction of fluvial and marine processes. For a delta to form there must be a heavily laden river, such as the Nile or the Mississippi, and a standing body of water with negligible currents, such as the Mediterranean or the Gulf of Mexico. Deposition is enhanced if the water is saline: since salty water causes small clay particles to flocculate or adhere together. Other factors include the type of sediment, local geology, sea-level changes, plant growth and human impact.

The material deposited as a delta can be divided into three types:

- **Bottomset beds** – the lower parts of the delta are built outwards along the sea floor by turbidity currents (currents of water loaded with material). These beds are composed of very fine material.
- **Foreset beds** – over the bottomset beds, inclined/sloping layers of coarse material are deposited. Each bed is deposited above and in front of the previous one, the material moving by rolling and saltation. Thus, the delta is built seaward.

- **Topset beds** – composed of fine material, they are really part of the continuation of the river's floodplain. These topset beds are extended and built up by the work of numerous distributaries (the main river has split into several smaller channels).

The character of any delta is influenced by the complex interaction of several variables:

- the rate of river deposition
- the rate of stabilisation by vegetation growth
- tidal currents
- the presence of longshore drift or not
- human activity (deltas often form prime farmland when drained).

There are many delta types, but the three 'classic' types are:

- **Arcuate delta** – fan-shaped. These are found in the areas where regular longshore drift of other currents keep the seaward edge of the delta trimmed and relatively smooth in shape, such as the Nile and Rhone deltas.
- **Cuspate delta** – pointed like a tooth or cusp, e.g. the Ebro and Tiber deltas, shaped by regular, but opposing, gentle water movement.
- **Bird's foot delta** – where the river brings down enormous amounts of fine silt, deposition can occur in a still sea area, along the edges of the distributaries for a very long distance offshore, such as the Mississippi delta.

Deltas can also be formed inland. When a rivers enters a lake it will deposit some or all of its load, so forming a **lacustrine delta**. As the delta builds up and out, it may ultimately fill the lake basin. The largest lacustrine deltas are those that are being built out into the Caspian Sea by the Volga, Ural, Kura and other rivers.

CASE STUDY: THE RHONE DELTA

The Rhone River divides into two main distributaries 4 km north of Arles. The east branch, the Grande Rhone, is the larger of the two, and carries 85% of the Rhone's water into the Mediterranean. At Arles the river is just 2 m above sea level and takes 47 km to reach the sea. The delta is criss-crossed by numerous small islands and abandoned channels, and active levees. Most of settlements and transport routes are located close to the river, where the land is slightly higher. Further away from the river, the land is lower, swampy and frequently covered with water. The same pattern exists along the west branch, the Petite Rhone.

Between these two limbs of the Rhone is a flat region, characterised by many flat marshes and lakes (etangs), known as the Camargue. The largest lake is the Etang de Vaccares, which is less than 1 m deep. The etangs receive most of their water from rainwater that becomes trapped between the slightly higher riverine locations, and the sand bars and dunes at the coast.

The Rhone delta is believed to be less than 1 million years old. Deposition by the Rhone is estimated to be about 17 m^3 each year, or about 50 tonnes every minute! As the Mediterranean Sea has a very low tidal range, there are no currents to carry away these deposits. In addition, the Mediterranean is very saline. In the presence of salt water clay and mud particles coagulate to form larger particles that cannot be held aloft by the flow of the river. Hence, there is rapid deposition at the mouth of the delta.

There are a number of stages in the formation of a delta. The first is the development of sandbanks in the original mouth of the river. This causes the river to divide, and then there is a period of repeated subdivisions until there is a large number of distributaries flowing towards the sea. Each of the channels develops its own set of levees, which has an impact on the human environment (affecting settlement and transport) as well as the physical environment, affecting the development of etangs between the main branches of the river. The etangs may slowly accumulate sediment to form marshes, or they may be drained and reclaimed to form farmland or other uses for people.

9 The River Course – Long and Cross Profiles

The long profile of a river is the gradient of the channel bed from source to mouth (Figure 21). Generally, river profiles are steep in their upper course and less steep in their lower course, producing a concave profile overall. Near the source of the river (the upper course) erosion is limited because the volume of water in the channel and the load are small. Erosion is also limited near the mouth (the lower course) because the river is heavily laden and much energy is expanded on transport. In a relative sense, therefore, erosion is at a maximum in the middle course. Thus, the graded profile of a river is typically concave.

In his *Cycle of Erosion*, WM Davis stated that a river's activity would change over time. In the early stages the river would be 'youthful' engaged primarily in cutting narrow, V-shaped valleys into the general land level. In the middle part of the time cycle, the river begins to 'mature', eroding and depositing material, while the action of weathering, etc., has begun to open up the valley sides, generally subduing the relief. In the last stages of the cycle, the river becomes more sluggish ('senile'), flowing over the land that has become degraded by other agents of weathering and erosion. These three stages are, according to Davis, reflected in the long and cross profiles of many rivers: the upper reaches reflect the 'youthful' stage, the middle reaches the 'mature' stage, while the lower reaches reflect the 'senile' stage.

In recent years, considerable criticism has been made of this concept of 'grade'. Such a graded profile may never be achieved by some rivers for a number of reasons. These include:

- the presence of local base levels
- changes in the general activity of the river by rejuvenation.

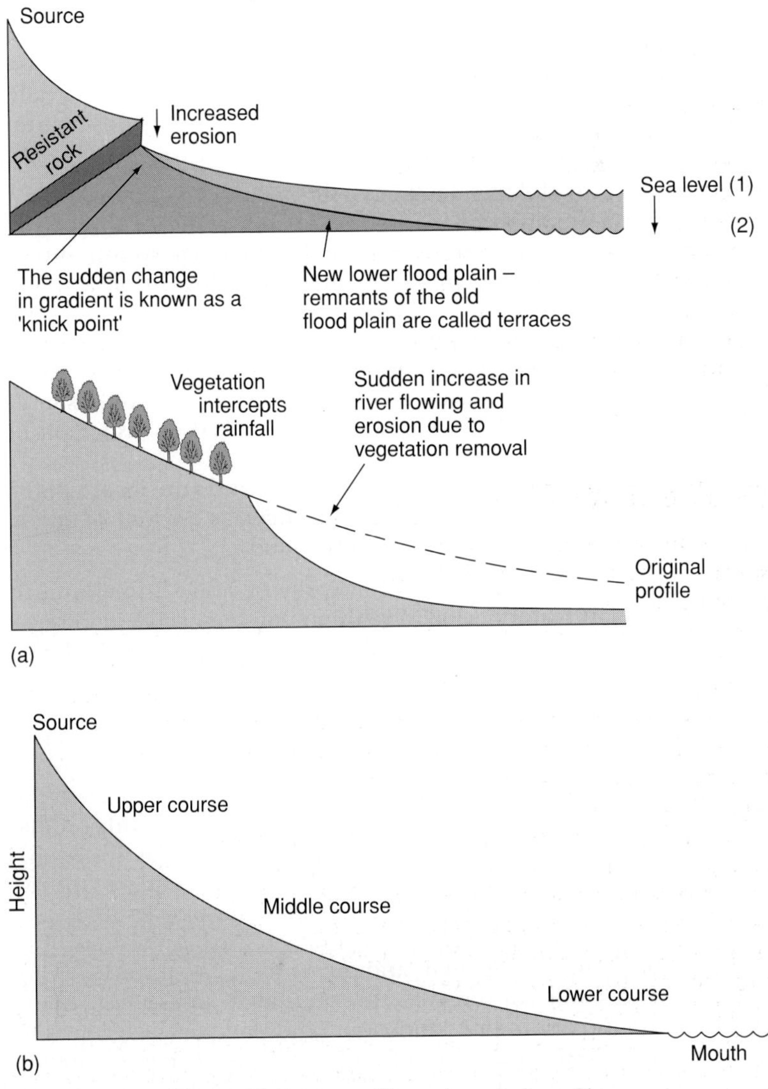

Figure 21 Long profile and graded profile

a) Local base levels

Waterfalls and lake basins form important local base levels. However, many of these are only temporary interruptions in the long profile. For example, waterfalls may retreat and lakes may fill. Sediment is deposited in the lake as a lacustrine delta that gradually enlarges to fill the lake. Meanwhile, the outflowing water will accelerate down the steeper slope downstream of the basin, causing headward erosion at the outlet. Both processes will cause the lake and local base level to disappear.

b) Rejuvenation

Rejuvenation can seriously interrupt the attainment of a graded profile (Figure 21a). It causes the river to increase its downcutting activity. Rejuvenation is caused by:

- tectonic activity (folding, faulting, uplift, subsidence, tilting) causing dynamic rejuvenation
- an increase in the volume of water in the drainage system, due to:
  (i) an increase in precipitation in the catchment area due to climatic change, or
  (ii) the result of river capture
- changes in base level (either sea level falls or the land rises) due to:
  (i) eustatic fall in sea level (a worldwide fall in sea level) as in the Pleistocene Ice Age when water was 'locked up' as ice on the land, or
  (ii) an isostatic change in the level of the land (the local uplift of the land, in late and post-glacial times as a result of the ice melting, relieving pressure on the land).

Rejuvenation has affected the vast majority of rivers throughout the world. As the last major period of rejuvenation has occurred in recent geological time, its effects can still be clearly seen in fluvial landscapes. Indeed, some areas that were heavily glaciated are still experiencing rejuvenation and landscape change.

Features associated with rejuvenation include knickpoints, river terraces, and incised meanders. With the fall in base level, the seaward end of the river will fall over a sharp change in gradient, creating a waterfall. The sudden change in gradient called a knickpoint. As with ordinary waterfalls, the plunging river will cause the river to retreat upstream. If stability continues for a long time, the knickpoint may retreat to the river source, creating a profile 'graded' to the new lower base level.

Rejuvenation causes the river to cut down, leaving portions of the former floodplain untouched and upstanding as terraces. These paired terraces will converge at the associated knickpoint. In one period of rejuvenation, one knickpoint and one pair of terraces will be created. However, there are often cycles of rejuvenation, creating a series of knickpoints and terraces. This is known as polycyclic relief (i.e. many cycles of rejuvenation).

Incised meanders are formed where the meander pattern of the river is maintained as the river increases its downcutting. There are two types of incised meander, ingrown and entrenched meanders. In an ingrown meander downcutting is slow. As the incision of the river continues, there is time for the ordinary processes of meander development and migration to operate. Hence, greatly enlarged river cliffs (bluffs) and slip-off slopes are created, and the meander has an asymmetric cross-section as with normal meanders. By contrast, entrenched meanders occur where downcutting is so rapid that meander migration is not allowed. Thus, the river cuts more of a winding gorge with a symmetric cross-section. Incised meanders are very well developed along the course of the River Wye.

Excellent examples of incised meanders can also be found on the River Dee near Llangolen. The Dee has cut into Silurian rock, west of Llangollen, to form incised meanders, while, to the east of Llangollen, the Dee has shortened its course by cutting through the neck of two meanders. Glacial debris has been found in the abandoned valleys, suggesting that the shortening (erosion) took place when the river was swollen by glacial meltwater, and thus had a considerably higher erosional potential.

As the long profile changes, so does the cross profile. The general characteristics of this cross-profile are controlled by:

- the type and rate of river activity
- the type and rate of weathering and downslope transport of this weathered material
- local geology.

c) River capture

Since both drainage basins on either side of a watershed are constantly at work potentially modifying the watershed by headward erosion, the watershed may migrate towards one or the other, eventually leading to breaching. The possibility of river capture will depend upon a number of factors:

- the relative power of erosion of the two systems
- the relative amount of precipitation in each of the basins (more precipitation leads to more energy)
- the geological structure of each basin and the resistance of the local rocks
- human activity such as deforestation in one catchment.

Some form of river capture is almost certain to occur anywhere, since it is highly unlikely that two adjacent systems will experience identical conditions. For one reason or another, one system will dominate over the other. River capture is common in Wales, for example the capture of the upper Teifi by first the Ystwyth and then the Rheidol.

CASE STUDIES: CLIFTON GORGE, SYMONDS YAT AND THE MISFIT RIVERS OF THE COTSWOLDS

Clifton Gorge

The Clifton Gorge near Bristol is cut by the River Avon through a plateau of Carboniferous limestone. Thus, the Avon flows from one lowland area to another through a deep gorge cut in a limestone escarpment. Recent research suggests that ice sheets moving over this area may have created meltwater channels which cut the gorge. A more likely explanation, however, is that the gorge at Clifton is an example of superimposed drainage. It is likely that the Carboniferous limestone of the area was covered by younger deposits of Jurassic and Cretaceous rocks. A drainage system developed on the younger rocks, eroded them and revealed the older rocks beneath. As the process was very slow, the rivers were not forced to change course, but merely maintained their course and began to cut down through the harder rocks, to form the gorge.

Symonds Yat

The meanders found at Symonds Yat are excellent examples of ingrown meanders. In the deep geological past, sea levels were up to 150 m above present and the River Wye would have developed a broad valley and floodplain, over 1.5 km wide. When the area around Symonds Yat was uplifted relative to sea level, the river was given renewed energy – rejuvenation – and the ability to erode down to the new base level. The Wye therefore began to erode downwards within its valley, producing a smaller valley within the larger floodplain. Downward erosion occurred relatively slowly since the uplift was slow and the underlying rocks relatively resistant. Hence, the river eroded laterally as well as vertically, and so meanders migrated downstream. Large river cliffs were created on the outside bends of the meanders while deposition continued on the inside banks. The Wye developed a highly asymmetric cross-section.

In contrast, where uplift is rapid, erosion tends to be vertical rather than lateral and so entrenched meanders are formed – symmetrical in cross-section and characterised by steep slopes on both sides of the meander.

The Wye is also an example of superimposed drainage, rather like the Avon at Clifton Gorge. It pays very little regard to the underlying geology or relief. At Symonds Yat, the river flows from high ground to low ground and back through high ground again crossing from Carboniferous limestone to sandstone and back again to limestone.

Misfit rivers

The Cotswolds are an escarpment of oolitic (Jurassic) limestone. Limestone is a permeable rock and therefore we would not expect to find much drainage or many valleys present. The steep scarp slope faces towards the Severn while the gentle dip slopes falls away towards the Thames Valley. Unusually, the Cotswolds contain many large valleys, yet they are occupied by small rivers such as the Windrush, Evenlode, Coln and Leach. These misfit rivers generally occupy large, steep-sided valleys that contain wide, flat bottoms.

It is thought that many of these valleys are the result of climate change. The channels associated with the valley meanders are generally 10 times as wide as the existing streams, and the wavelengths of the valley meanders are approximately nine times as long as the wavelength of the stream meanders.

During the cold, glacial phases, much of the precipitation in the region would have fallen as snow. Rapid spring melt would have created discharges up to 50 times greater than in the streams today. Such large amounts of runoff, combined with the frozen subsurface (permafrost) which would have reduced infiltration, would lead to increased overland runoff and stream erosion.

Summary

- River regimes show the average annual flow of water in a river. The main factor influencing this is climate.
- Storm hydrographs operate on a short-term basis – usually less than one week. These are influenced by a number of factors such as rainfall intensity, impermeability of the surface, vegetation cover, gradient and saturation of the soil.
- Human activities, such as urbanisation, are changing flood hydrographs.
- Water flow in streams is highly variable.
- Rates of erosion, and types of erosion, vary widely.
- Features of erosion include waterfalls, gorges and potholes.
- Features of deposition include levees and deltas.
- Some features can be considered the result of both erosion and deposition, such as meanders and ox-bow lakes.
- Rivers also vary in their long and cross profile. In addition, profiles vary over time.

Questions

1. Study Figures 10 and 11, which show the impact of urbanisation on flood hydrographs. Describe and explain the differences in the relationship between discharge and time.
2. Define the following terms: avulsion, crevasse splaying and distributaries.
3. Why does flocculation lead to increased deposition of clay?
4. Why are deltas (**i**) attractive places to live and (**ii**) hazardous places to live? Illustrate your answer with examples.
5. Using an atlas, explain why the Nile and the Mississippi are heavily laden with sediments.
6. Explain briefly the life cycle of a delta under natural conditions. What was the average annual growth rate for the distributaries in the West Bay part of the Mississippi Delta? Why is silt deposited first rather than the clay?

3 Wetlands

KEY DEFINITIONS

Bog A peat-accumulating wetland that has no significant inflows or outflows and supports mosses, particularly sphagnum
Fen A peat-accumulating wetland that receives some drainage from surrounding mineral soil and usually supports marsh-like vegetation
Marsh A frequently or continually inundated wetland characterised by emergent herbaceous vegetation adapted to saturated soil conditions. In European terminology, a marsh has a mineral soil substrate and does not accumulate peat.
Peatland A generic term for any wetland that accumulates partly decayed plant matter.
Swamp In the US, a wetland dominated by trees or shrubs. In Europe, a forested fen would easily be called a swamp. In some areas, wetlands dominated by reed grass are also called swamps.
Wetland Land with soils that are permanently flooded.

1 Introduction

There is a wide variety of wetland habitats. These account for about 6% of the world's habitats, and provide an important resource for those who live on them. However, wetlands have not always been viewed favourably, and for much of the 20th century they have been destroyed, altered, drained and removed to make way for agriculture, settlements, transport and industrial developments. Wetlands are under increasing pressure as a result of global warming. Nevertheless, examples of sustainable management of wetlands show that it is possible at a local level and at a regional/international level to manage wetlands sustainably.

2 Wetlands

A wetland is defined as 'land with soils that are permanently flooded'. The Ramsar Convention, an international treaty to conserve wetlands, defines wetlands as 'areas of marsh, fen, peatland or water, whether natural or artificial, permanent or temporary, with water that is static or flowing, fresh, brackish or salt'. Thus, according to the Ramsar classification, there are marine, coastal, inland and man-made types, subdivided into 30 categories of natural wetland and nine human-made ones, such as reservoirs, barrages, and gravel pits. Wetlands now represent only 6% of the Earth's surface, of which 30% are bogs, 26% are fens, 20% are swamps, 15% are floodplains and 2% are lakes. It is

Figure 22 Wetlands provide a range of functions – ecological, environmental and recreational

estimated that there was twice as much wetland area in 1900 compared with 2000.

Wetlands provide many important social, economic and environmental benefits (Figure 22). These include water storage, groundwater recharge, storm protection, flood mitigation, shoreline stabilisation, erosion control, and retention of carbon, nutrients, sediments and pollutants. Wetlands also produce goods that have a significant economic value such as clean water, opportunities for tourism, fisheries, timber, peat and wildlife resources. The loss and degradation of wetlands is caused by several factors including:

- increased demand for agricultural land
- population growth
- infrastructure development
- river flow regulation
- invasion of non-native species and pollution.

Wetland functions can generally be grouped into three main types: regulation, provision of habitats and production. Wetlands are important regulators of water quantity and water quality. Floodplain wetlands, for example, store water when rivers over-top their banks, reducing flood risk downstream. Wetlands also regulate water quality. Reedbeds and other wetland plants, for example, are known as important regulators since they remove toxins and excessive nutrients from the water.

Wetlands are characterised by a large number of ecological niches and are very diverse. Many components of wetland ecosystems also provide resources for direct human consumption including water for drinking, fish, reeds for thatched roofs, timber for construction, peat and fuelwood. Recreational uses include fishing, sport hunting, bird-watching, photography and water sports.

Wetlands are highly dependent on water levels and so changes in climatic conditions that affect water availability, such as long-term increases in temperature, sea-level rise and changes in precipitation, will highly influence the nature and function of specific wetlands. The ability of wetland ecosystems to adapt will be highly dependent on the rate and extent of these changes. Nevertheless, there is uncertainty about the increase in frequency and intensity of extreme events, such as storms, droughts and floods.

CASE STUDY: THE RAMSAR CONVENTION (http://iucn.org/themes/wetlands/)

The Convention on Wetlands of International Importance Especially as Waterfowl Habitat (usually referred to as the Ramsar Convention after the place of its ratification in Iran in 1971) is one of the most important instruments for conserving wetlands. This treaty laid the basis for international cooperation in conserving wetlands and more than 60 countries have signed up to the Ramsar Convention.

The convention requires the signatories:

1 To designate wetlands of international importance for inclusion in a list of so-called Ramsar sites.
2 To maintain the ecological character of their listed Ramsar sites.
3 To organise their planning so as to achieve the wise use of all of the wetlands on their territory.
4 To designate wetlands as nature reserves.

There are more than 500 wetland sites on the Ramsar list covering in excess of 30 million hectares of wetland habitat. To be considered a wetland of international importance, a site must:

- support a significant population of waterfowl, threatened species, or peculiar fauna or flora;
- be a regionally representative example of a type of wetland;
- be physically and administratively capable of benefiting from protection and management measures.

A distinguishing characteristic of the Ramsar Convention on Wetlands is the adoption of the concept of 'wise use' as a part of the idea of nature conservation. The wise use of wetlands is their sustainable utilisation for the benefit of mankind in a way compatible with the maintenance of the natural properties of the ecosystem. In this context, 'sustainable utilisation' is defined as 'human use of a wetland so that it may yield the greatest continuous benefit to present generations while maintaining its potential to meet the needs and aspirations of future generations'.

3 Wetland Conservation

The perception of wetlands has not always been favourable. Historically, wetlands were considered hazardous, marginal waterlogged lands, harbouring disease. In the past their value was largely ignored and so the conservation of wetlands was not considered to be important. However, it is increasingly realised that wetlands are important and that they need to be valued (Figure 23). Wetlands are one of the most productive ecosystems in the world, and the total economic value (TEV) of wetlands has often been underestimated. This TEV includes:

- direct use values of products such as fish and fuelwood, and services, such as recreation and transport
- indirect use values such as flood control and storm protection provided by mangroves
- option values, which could be discovered in the future
- intrinsic values, the value of the wetland ' of its own right' with its attributes.

A study of the large Hadejia-Nguru wetland in arid northern Nigeria found that water in the wetland yielded a profit in fish, firewood, cattle grazing lands and natural crop irrigation that was 30 times greater than the yield of water being diverted from the wetland into a costly

Functions	Products	Attributes
Flood control	Fisheries	Biological diversity
Sediment accretion and deposition	Game	Culture and heritage
Groundwater recharge	Forage	
Groundwater discharge	Timber	
Water purification	Water	
Storage of organic matter		
Food-chain support/cycling		
Water transport		
Tourism/recreation		

Figure 23 The value of wetlands

irrigation project. A recent attempt to put a dollar value on the ecological services provided by different ecosystems worldwide put wetlands top at almost US$15,000 per hectare per year, seven times that of tropical rainforest. Much of this value comes from flood prevention.

Natural causes of wetland loss include sea-level rise, drought, hurricanes and other storms, and erosion. However, the main cause has been human interference, either direct or indirect. The reclamation of wetlands and dam construction cause the greatest loss of wetland habitat. Other causes include the conversion for aquaculture, mining of wetlands for peat, coal, gravel, phosphate and other materials, and groundwater abstraction. In other cases it has been down to political hostilities.

CASE STUDY: DRAINAGE OF THE IRAQI MARSHES

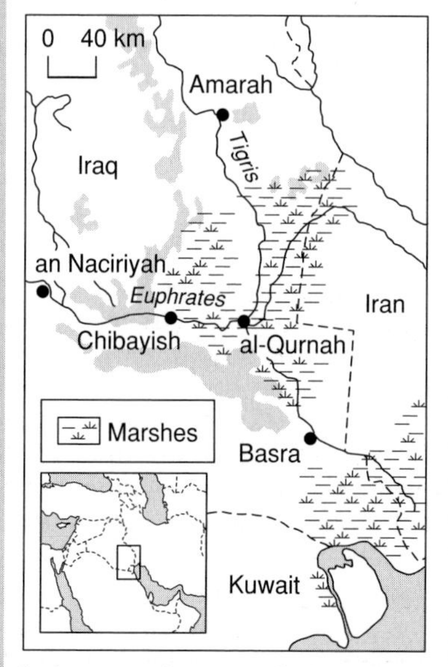

The village of Chibayish typifies much of former marshes in the south-east of Iraq. The village, once an island surrounded by marshes, is empty and in ruins. Chibayish was once home to the Madan people, the Shia Muslims who lived on the broad marshes that stretched across southern Iraq. During the Gulf war (1991), the villagers of Chibayish, like hundreds of thousands of Shias in the south, rose up against Saddam Hussein in a vast but eventually futile rebellion.

During the past decade the systematic draining of the marshlands north of Basra, has led to massive population displacement. Massive irrigation works have dried the wetlands, decimated the population and destroyed a way of life which had lasted for 5000 years.

Most of the original marsh Arabs, who in 1991 numbered around 250,000, have become displaced. The population of the marshes has been completely decimated.

From the early 1950s engineers in Turkey, Syria, Iran and Iraq began building a series of huge dams across the Euphrates and Tigris rivers. In the early 1990s President Saddam's engineers dried the marshes. In their place are long, straight canals.

(Adapted from *The Guardian*, 6 January 2003)

There are many problems facing those responsible for the conservation of wetlands into the 21st century:

- the prevention of further damage and loss of wetland habitat;
- conserving remaining wetlands
- rehabilitating damaged wetlands
- creating new wetlands.

An excellent example of sensitive management is Wicken Fen in East Anglia, UK. It is located approximately 15 km north-east of Cambridge. This fen covered approximately 3380 km^2 in the 17th century but was largely drained in two phases between 1630 and 1653. Between 1637 and 1954, there was a reduction in area of the East Anglian fens from 3380 km^2 to 10 km^2.

Wicken Fen is one of the oldest nature reserves in Britain, and since 1899 it has been owned and managed by the National Trust as a remnant of a once extensive landscape. Wicken Fen is a Ramsar site, a National Nature Reserve, and a Site of Special Scientific Interest. Wicken Fen was a summer dry/winter wet fen during the 1630s and it had various traditional uses: sedge for thatching, peat cutting for fuel and 'litter' (common reed and purple moor-grass) for animal bedding. The underlying clay was used for local brick making. Wild crops of reed and sedge are still harvested, and Wicken Fen attracts 30,000 visitors a year.

The reserve is about 305 ha and consists of four sections. The most important section – Wicken Sedge Fen (103 ha) – is kept wet by pumping water into it, and by waterproofing its perimeter. The main problems for Wicken Fen's conservation have been falling water levels and encroachment by scrub. The Fens would quickly turn into woodland if not continuously wet, and in fact much of Wicken Fen converted to wood in the 1950s. In 1961 a management plan was drawn up to arrest the fen's decline and to restore its former habitats. Management techniques have included, for example, the reopening old ditches, excavating new ditches, cutting down trees and reintroducing sedge harvesting. In 1982, the fen violet reappeared – 66 years after it had last been seen at the site.

Thus, at Wicken Fen, there is a small-scale example of agricultural land reverting to nature, a perfect example of wetland restoration.

CASE STUDY: THE OKAVANGO BASIN

The Okavango Basin is the fourth largest international river basin in southern Africa, covering up to 721,277 km^2. The main river in the catchment is the Okavango, which rises in the southern Angolan highlands, flows south-eastwards to Namibia and then to Botswana, where it supplies the Okavango Delta (Figure 24). The Okavango Delta is a large inland delta or alluvial fan consisting of

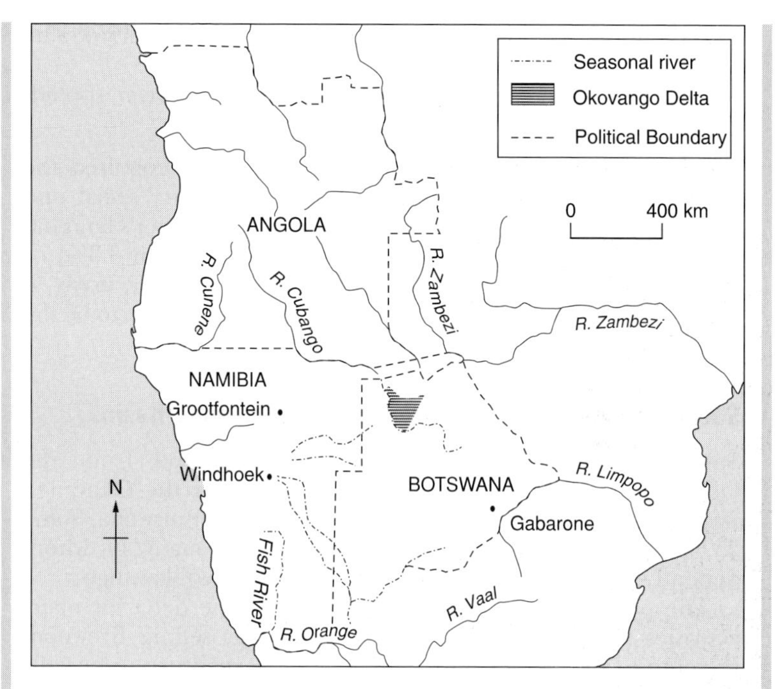

Figure 24 Okavango Delta

about 6000 km² of permanent swamp, and between 7000 km² and 12,000 km² of seasonally inundated swampland. Sometimes referred to as 'the jewel of the Kalahari' the Okavango Delta is the only Ramsar site in the Okavango Basin, and is the largest Ramsar site in the world. In September 1994, Angola, Namibia and Botswana signed an agreement to create the Okavango River Basin Water Commission (OKACOM).

Water inflow to the Okavango Delta averages 10 million m³ per annum, ranging from 7 million m³ in dry years to 15 million m³ in wet years. Most of the inflow (97%) is lost to evapotranspiration and seepage. The Okavango River flows mainly during the summer months and its floods supply the waterways, reedbeds, floodplain and islands that make up the Okavango Delta in Botswana.

The Okavango catchment can be divided roughly into three different zones:

- the Angolan region, which contains numerous tributaries which feed into the river – where there are a number of districts that have a high potential for irrigated and flood-recession agriculture

- the middle section, in which the river flows in a narrow alluvial plain up to 6 km wide
- the Panhandle region in Botswana, where the river spreads out eventually into the Okavango Delta.

The North-west District of Botswana has long recognised the importance of natural resources to its economic, social and political well-being. A sizable proportion of the district's land has been designated either as game reserve (3600 km^2, or 3.3%) or Wildlife Management Areas (19,100 km^2, or 17.5%), areas in which wildlife and habitat conservation and tourism are to be the primary land uses.

Socioeconomic characteristics of the Okavango Basin

Some 100,000 Namibians gain their livelihood from the Okavango River. Overall population density for the Okavango Basin is low, approximately 3 people/km^2. In Botswana, some 25,000 people live in the Okavango Delta, many of whom depend on the water and other resources of the Okavango.

The people of the Okavango region use the delta for many activities such as foraging; hunting; fishing; selling firewood; thatching grass, poles and palm leaves; agriculture, especially flood-recession farming (molapo); livestock-raising; and wage labour. The delta area contains a range of resources such as water, wild plant foods and medicines, wild animals, trees and shrubs for shelter, fuel and building materials, and materials such as termite earth and ochre used for construction and for adornment. African river basin and inland delta ecosystems are highly productive. They generally contain rich soils, relatively abundant fish populations, and sufficient water to allow flood water farming and irrigation. As a result, they have been a source of intense interest to those who wish to enhance economic development opportunities.

Development plans

The Okavango Delta and the Okavango River have long been of interest to planners and government development agencies in southern Africa, especially for irrigation schemes, water transfer scheme and hydroelectric power.

In the 1980s the Botswana government proposed the Southern Okavango Integrated Water Development Project (SOIWDP). This project was opposed strongly by local people. The Botswana government agreed to an independent review of the project, a major precedent, since it was the first time that a national government had asked an outside agency to conduct a review of a major water development project. A set of alternatives provided by the IUCN consultants emphasised the use of groundwater, the diversification of local economies and community involvement.

In 1996 the Namibian government decided to extract water from the Okavango and to transfer the water by pipeline to the Eastern National Water Carrier at Grootfontein which would, in turn, transfer water to the Windhoek area in central Namibia. The initial proposal of the Namibian government was to do an Environmental Impact Assessment (EIA) only in Namibia. However, the Botswana government, and various environmental organisations, pointed out that the EIA should examine downstream impacts of the water extraction project as well. It concluded that the delta was a major wetland that supports sizable human and wildlife populations, and that it is an important tourist destination.

Water use in Botswana

Water use in Botswana was estimated in the mid-1980s to be approximately 350,000 m^3 per capita per day. Irrigation and livestock were the main water users in National Development Plan 6 (1985–91) (irrigation: 35.1%, livestock, 35.6%, urban, 11.6%, mining, 12.6%, and villages and other uses, 5.1%). Urban and village water use has increased quickly as a result of population growth and economic development, as well as urbanisation and better infrastructure. The increase in water demand and use has been greatest in Gaborone, where the water consumption in 1985 was seven times what it was at independence (1966).

Surface water sources in Botswana such as the Okavango Delta generally have a number of characteristics:

- the largest sources occur where demand is presently low
- the catchment areas of most large sources are partly outside the country
- the water bodies have high rates of evapotranspiration (up to 2 m or more per annum)
- the availability of water is related positively to rainfall, which is highly variable in space and time, and the size of the catchment area.

There are many government organisations interested in water development in Botswana, and there are many NGOs too. These include the Kalahari Conservation Society, an NGO mainly involved with wildlife conservation but it also deals with integrated resources management and conservation education; and Permaculture, an NGO that promotes sustainable agriculture and resource management in rural areas

Coping with drought in the Okavango Basin

An almost universal response of African peoples to drought is to diversify their strategies, exploiting a variety of different kinds of resources and taking advantage of economic opportunities. Another common response is to move, either away from places afflicted by drought or to places where resources are available. Other strategies include the formation of sharing and exchange systems, kinship and social alliances, and food storage.

The serious droughts of 1982–7 and 1992–3 created many problems for local people. Large numbers of livestock died, thus reducing the chances of rural households for gaining cash through sales of cattle or small stock. Wild plant foods and game were also depleted seriously, thus reducing opportunities for people to use resources that often served as fallback goods in stress periods.

One of the strategies for coping with drought and climatic uncertainty employed by local people was to request permission to move to another group's territory, which had sufficient resources to sustain a larger number of people. Usually people asked those with whom they already had social contacts.

Sustainable community-based development

Sustainable development has been defined by the World Commission on Environment and Development as that which '... meets the needs and aspirations of the present without compromising the ability of future generations to meet their own needs'. A number of efforts have been made in the Okavango Basin to allow communities to assume control over natural resources, particularly wildlife.

For example, changes in land use have occurred in the Okavango region, particularly with the imposition of wildlife conservation laws and the establishment of the Moremi Game Reserve, one of the first tribal game reserves in Africa. There were also changes in land management and administration patterns, especially after Botswana's independence in 1966. The powers of traditional authorities (chiefs and ward heads) over land allocation were transferred to government land boards under the Tribal Land Act of 1968. Vegetation resources are covered in part by the Agricultural Resources Conservation Act and the Herbage Preservation (Prevention of Fires) Act, and a range of conservation activities are promoted through the Agricultural Resources Board (ARB) of Botswana.

Fishing activities in the Okavango region are controlled to a limited extent by the Fisheries Unit in the Ministry of Agriculture. The Forestry Unit in the Ministry of Agriculture engages in efforts to promote the sustainable use and conservation of timber resources.

There are differences of opinion over the future of the Okavango Delta between non-government organisations, international agencies, and the governments of Botswana and Namibia as well as other governments in the southern African region. This struggle, in the eyes of some, is over conservation versus development. In the eyes of others, the struggle is for sustainable development that allows for use of resources by present-day generations, with an eye toward ensuring the viability of the ecosystems over the long term. It does appear at present that both Botswana and Namibia have tended to place state and private interests above those of local communities. One of the trends in the Okavango region in Namibia and Botswana is toward greater privatisation. In Botswana, dozens of safari camps have been established in the delta.

4 Wetlands and Climate Change

Wetlands and peatlands represent important carbon stores and contribute significantly to the global carbon cycle. According to the IPCC (1996) since pre-industrial times, sea levels have risen globally between 1 and 5 m. Continuing sea-level rise during the 21st century would double the global population at risk from storm surges from around 45 million up to 90 million. The most vulnerable areas are small island states and low-lying areas such as Bangladesh and other states in south-east Asia, north-western Europe, the southern Atlantic coast and the Gulf of Mexico in the USA. The effects of an increase of coastal erosion include increased coastal flooding, loss of habitats, an increase in the salinity of estuaries and freshwater aquifers, and changed tidal ranges in rivers and bays, transport of sediments and nutrients, and patterns of contamination in coastal areas are amongst the main effects of coastal erosion.

Coastal wetland flora and fauna usually respond to small, permanent changes in water levels. However, the degree to which they are able to adapt to these changes will depend to a great extent on the ability for species to 'migrate' to other areas. Many coastal and estuarine wetlands will be unable to migrate inland due to the presence of dykes, levees or specific human land uses close to the coastal area, such as urban, industrial and transport developments.

Higher sea levels and increased storm surges could also adversely affect freshwater supplies available from coastal wetlands due to saltwater intrusion. Salt water in delta systems would advance inland affecting the water quality available for agricultural, domestic and industrial use. The Nile Delta is a good example of the potential problems.

Estimates of the loss of wetland in industrialised regions indicate that up to 60% of these have been destroyed in the last 100 years due to drainage, conversion, infrastructure development and pollution. These changes are believed to be responsible for most of the loss in freshwater biodiversity in the USA in recent decades.

The situation is complex. For example, water demand is projected to increase steadily during the coming decades, while water availability, especially in arid and semi-arid areas, is likely to decrease. To meet this problem, many countries will need to increase reservoir storage capacity (dam building) to meet the increasing demands for irrigation. This could have an adverse impact on wetlands. For example, using HEP as an alternative to fossil fuel would lead to more dam construction. In China, dam construction for hydropower is already expected to increase by 6% annually. The construction of dams will put additional stress on wetland ecosystems by increasing habitat fragmentation. Dams also retain large quantities of sediments essential to the maintenance of deltas and coastal wetlands.

By increasing water storage capacity through dam construction there will be an increase in the build-up of silt in reservoirs. This will cause further coastal and delta erosion. Gradual compaction of delta soils and delta wetlands will cause subsidence, causing the delta to fall below sea level. The combination of sea-level rise and land subsidence could place human populations in the deltas and the coastal zone at additional risk.

Although wetlands only cover about 8–10% of the world's land surface, they contain 10–20% of the global terrestrial carbon. They therefore play an important role in the global carbon cycle. The carbon pool contained in wetlands is estimated to amount up to 230 gigatonnes (Gt) out of a total of about 1943 Gt. Peat deposits are estimated to hold 541 Gt of carbon in total.

Figure 25 summarises greenhouse gas stores and flows in wetlands. Figure 25(a) shows the significant amount of carbon stored in peatland soils, particularly in tropical peatland soils, and biomass. Figure 25(b) shows the considerable amount of carbon dioxide (CO_2) emissions from swamps and bogs due to drainage and conversion to agriculture.

To assess properly the potential of natural wetlands in global warming the flows of CO_2, methane (CH_4) and nitrous oxide (N_2O) have to be taken into account. Wetlands and rice fields account for up to 40% of the global source of methane emissions to the atmosphere. When wetlands are converted to agricultural land, large quantities of CO_2 and N_2O (nitrous oxide) are released, although methane emissions are sharply reduced. Wetlands in northern Europe accumulate between 0.16 and 0.25 t C/ha/year but if methane emissions are taken into account these wetlands become a net source of 0.43–1.1 t C/ha/year.

(a) Carbon stocks and flows of peatlands

	Carbon stores (t C ha^{-1})	(t C ha^{-1})	Carbon absorption (t C/ha/year)
	Soil	Biomass	
Global	1181–1537	No data	0.1–0.35
Tropics	1700–2880	500	No data
Boreal/temperate regions	1314–1315	120	0.17–0.29

(b) CO_2 emissions from the conversion of wetlands (swamps and bogs only)

	CO_2 emissions	
	Drainage (t C/ha/year)	Agricultural use (t C/ha/year)
Global	0.23–0.26	1–10
Boreal/temperate regions	0.1–0.32	1–19

Figure 25 Greenhouse gas stores and flows in wetlands

5 Coping with Change

Wetlands store large amounts of carbon and when these wetlands are lost or degraded, CO_2 and other greenhouse gases are released into the atmosphere in large quantities. Therefore, conserving wetlands is a viable way of maintaining existing carbon stores and avoiding CO_2 and other emissions. Adaptation in the context of climate change can be defined as 'a deliberate management strategy to minimise the adverse effects of climate change, to enhance the resilience of vulnerable systems, and to reduce the risk of damage to human and ecological systems from changes in climate'. Wetland rehabilitation and restoration can be a viable alternative to flood control and dredging designed to cope with larger and more frequent floods.

The effectiveness of reservoirs to reduce peak discharges that might increase due to climate change remains doubtful. In many cases, reservoirs are often entirely filled by the time an extreme flood event occurs, meaning they have no storage capacity to reduce the flood peak in height or duration, as happened during the 2000 floods in Mozambique (see *Climate and Society* in the Access to Geography series).

The construction of levees and dykes, and straightening of channels to enable rapid drainage, are increasingly seen as counter-productive management interventions. An alternative management strategy is the restoration and rehabilitation of riverine wetland areas to enable large areas of land to be flooded. In response to the 1992 and 1993 floods, the Netherlands, for example, has started floodplain rehabilitation along the River Rhine to allow controlled flooding of areas during extreme discharges. In Napa Valley, California, the US Army Corps of Engineers have recently started with the development of alternatives to traditional flood prevention infrastructure, focusing partly on wetland restoration.

Summary

- Wetlands are critically important ecosystems, providing significant social, economic and ecological benefits such as regulation of water quantity and quality; habitat for waterfowl, fish and amphibians; resources to meet human needs; recreation and tourism.
- Climate change will degrade these benefits
- Climate change may lead to shifts in the geographical distribution of wetlands.
- Climate change will affect wetlands through sea-level rise; increased sea temperatures; changes in hydrology; increased temperature of wetland water bodies; increased temperature in tundra and polar areas
- Sea-level rise and increases in storm surges could result in erosion of shores and habitat; increased salinity of estuaries and freshwater aquifers; altered tidal ranges in rivers and bays; changes in sediment and nutrient transport; increased coastal flooding; increased vulnerability of some coastal populations.
- Land-use change and water consumption patterns will accentuate climate change impacts on wetlands.
- Although wetlands cover only a small portion of the world's land surface, they are significant carbon stores globally.
- Conversion and degradation of wetlands releases carbon and methane into the atmosphere in large quantities
- Conserving, maintaining and restoring wetlands avoids human-induced greenhouse gas emissions.

Questions

1. With the use of examples, describe some of the benefits of wetlands.
2. Suggest contrasting reasons why wetlands are being destroyed.
3. In what ways are wetlands vulnerable to climate change?
4. Outline, with the use of examples, ways in which wetlands can be managed sustainably.

4 Floods and Hard Engineering Structures

KEY DEFINITIONS

Bankfull discharge The discharge measured when a river is at bankfull stage.
Bankfull A condition in which a rivers channel fills completely, so that any further increase in discharge results in water overflowing the banks.
Channel The passageway in which a river flows.
Channelisation Modifications to river channels, consisting of some combination of straightening, deepening, widening, clearing or lining of the natural channel.
Discharge The quantity of water that passes a given point on the bank of a river within a given interval of time – usually expressed in cumecs – cubic metres per second.
Flash flood A flood in which the lag time is exceptionally short, i.e. hours or minutes.
Flood A discharge great enough to cause a body of water to overflow its channel and submerge surrounding land.
Flood-frequency curve Flood magnitudes that are plotted with respect to the recurrence interval calculated for a flood of that magnitude at a given location.
Floodplain The part of any stream valley that is inundated during floods.
Natural hazards The wide range of natural circumstances, materials, processes and events that are hazardous to humans, such as locust infestations, wildfires or tornadoes, in addition to strictly geological hazards.

1 Introduction

In this chapter we look at the causes of floods in rivers, and some of the consequences. (For detailed case studies of recent flooding see also *Climate and Society* in the Access to Geography series.) The causes of floods can be natural, however, human interference intensifies many floods. The decision to live in a floodplain, for a variety of perceived benefits, is one that is fraught with difficulties. The increase in flood damage is related to the increasing number of people living in floodplain regions.

Traditionally, floods have been managed by methods of 'hard engineering'. This largely means dams, levees and straight channels, which are wider and deeper than the ones they replace. Although hard engineering may reduce floods in some locations, they may cause unexpected effects elsewhere in the drainage basin. The case studies of the Mississippi and the Colorado illustrate this clearly.

2 Floods

Floods are one of the most common of all environmental hazards (Figure 26). This is because so many people live in fertile river valleys and in low-lying coastal areas. However, the nature and scale of flooding varies greatly. Less than 2% of the population of England and Wales and in Australia live in areas exposed to flooding. In contrast, 10% of the US population live within the 1:100 year flood. The worst problems occur in Asia where floods damage about 4 million hectares of land of each year and affect the lives of over 17 million people. Worst of all is China where over 5 million people have been killed in floods since 1860.

A key characteristic of water is its extreme events: floods and droughts. Between 1973 and 1997 an average of 66 million people a year suffer flood damage. This makes flooding the most damaging of all natural disasters (including earthquakes and drought). The average annual number of flood victims jumped from 19 million to 131 million in 1993–7. In 1998 the death toll from floods hit almost 30,000. The economic losses from the great floods of the 1990s are 10 times those of the 1960s in real terms. In addition, the number of disasters has increased by a factor of five.

Floods:

- account for about one-third of natural catastrophes
- cause more than half the fatalities
- are responsible for one-third of the economic losses
- have less than a 10% share in insured losses.

There are a number of reasons for the increase in the number of catastrophes and in the amount of damage they cause:

- population trends globally and in exposed regions
- increase in exposed values
- increase in the vulnerability of structures, goods and infrastructure
- construction in flood-prone areas
- failure of flood protection systems
- changes in environmental conditions – for example, clearance of trees and other vegetation and infilling of wetlands that reduces flood-retention capacities.

The major flooding in 2002 on the Danube, Elbe and Moldova in Europe were the worst floods in Europe for centuries, possibly since the Millennium floods in 1342. They caused about 300 deaths and $18.5 billion damage.

Some environments are more at risk than others. The most vulnerable include the following:

- Low-lying parts of active floodplains and river estuaries, for example in the lower Thames in London or in Bangladesh where 110 million people live relatively unprotected on the floodplain of

The 10 largest natural catastrophes 2002

Date	Country	Event	Fatalities	Economic loss	Insured loss ($ million)
Ranked by economic loss					
August	Europe	Floods	230	18,500	3,000
Aug-Sept	N & S Korea	Typhoon Rusa	150	4500	170
Jul-Aug	USA, Nebraska	Drought, heatwaves	3300		
June	China	Floods	500	3,100	
July-Dec	Australia	Drought		3,000	
September	West-central Europe	Storm Jeanette	33	2,300	1,500
September	Caribbean, USA	Hurricane Lili	8	2000	750
August	China	Floods	250	1,700	
September	France	Floods	23	1,200	450
April–May	USA	Tornadoes	10	2,000	1,500
Ranked by insured loss					
August	Europe	Floods	230	18,500	3,000
September	West-central Europe	Storm Jeanette	33	2,300	1,500
April-May	USA	Tornadoes	10	2,000	1,500
Sept	Caribbean, USA	Hurricane Lili	8	2000	750
November	USA	Tornadoes		600	460
September	France	Floods	23	1,200	450
Jan-Feb	USA	Winter storm	28	400	300
February	Germany, UK	Winter storm	3	500	300
September	Central America, USA, Caribbean	Hurricane Isidore	15	850	250
March	USA	Severe storm		250	210
Ranked by fatalities					
March	Afghanistan	Earthquake	2,000		
July-Aug	Bangladesh, India, Nepal	Floods	1,200	80	
May	India	Heat wave	1,100		
June	China	Floods	500	3,100	
May	Bangladesh	Severe storm	270		
August	China	Floods	250	1,700	
June	Iran	Earthquake	245	300	
August	Europe	Floods	230	18,500	3,000
July-Oct	Vietnam	Floods	155	25	
Jan-Feb	Indonesia	Floods	150	350	200

Figure 26 Floods and other hazards

the Ganges, Brahmaputra and Meghna. Floods caused by the monsoon regularly cover 20–30% of the flat delta. In very high floods up to half of the country may be flooded. In 1988 46% of the land was flooded and over 1500 people were killed.

- Small basins subject to flash floods. These are especially common in arid and semi-arid areas. In tropical areas some 90% of lives lost through drowning are the result of intense rainfall on steep slopes.
- Areas below unsafe or inadequate dams. In the USA there are about 30,000 sizable dams and 2000 communities are at risk from dams.
- Low-lying inland shorelines such as along the Great Lakes and the Great Salt Lake in the USA.
- Alluvial fans in semi-arid areas are prone to flash floods.

In most developed countries the number of deaths from floods is declining, although the number of deaths from flash floods is changing very little. By contrast, the average national flood damage has been increasing. The death rate in developing countries is much greater, partly because warning systems and evacuation plans are inadequate. It is likely that the hazard in developing countries will increase over time rather than decrease as more people migrate and settle in low-lying areas and river basins. Often newer migrants are forced into the more hazardous zones.

Since World War II there has been a change in the understanding of the flood hazard, in the attitude towards floods and policy towards

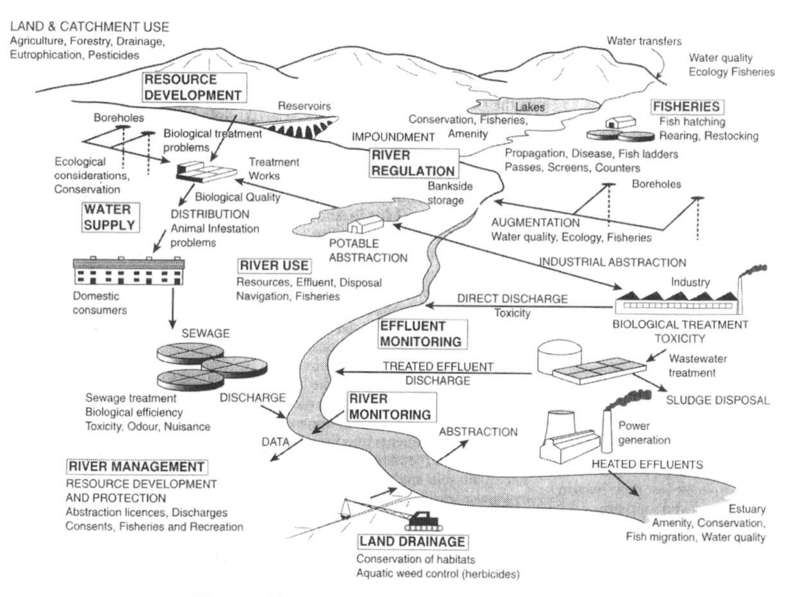

Figure 27 Hard engineering structures

reducing the flood hazard. The focus of attention has shifted away from physical control (engineering structures) towards reducing vulnerability through non-structural approaches. Three overlapping stages have been identified:

- the structural era, 1930s to 1960s (hard engineering, namely reservoirs, levees, channel improvements (Figure 27))
- the unified floodplain management era, 1960s to 1980s (flood warning, land use planning, insurance)
- post-flood hazard mitigation era, 1980s onwards (property acquisition and land-use control).

3 Physical Causes of Floods

A flood is a high flow of water which over-tops the bank of a river. The primary cause of floods are mainly the result of external climatic forces, whereas the secondary flood intensifying conditions tend to be drainage basin specific. Most floods in Britain are associated with deep depressions (low-pressure systems) in autumn and winter, which are both long in duration and wide in areal coverage. The floods in the UK during the winters 2000–1 and 2002–3 are good examples. By contrast, in India, up to 70% of the annual rainfall occurs in 100 days in the summer south-west monsoon. Elsewhere, melting snow is also responsible for widespread flooding.

Flood-intensifying conditions cover a range of factors that alter the drainage-basin response to a given storm. These factors include topography, vegetation, soil type, rock type and characteristics of the drainage basin (Figure 28).

The potential for damage by flood waters increases exponentially with velocity and speeds above 3 m/s and can undermine the foundations of buildings. The physical stresses on buildings are increased even more, probably by hundreds of times, when rough rapidly flowing water contains debris such as rock, sediment, debris and trees. Other conditions that intensify floods include changes in land use. Urbanisation, for example, increases the magnitude and frequency of floods in at least four ways:

- creation of highly impermeable surfaces, such as roads, roofs and pavements
- smooth surfaces served with a dense network of drains, gutters and underground sewer increase drainage density
- natural river channels are often constricted by bridge supports or riverside facilities reducing their carrying capacity
- due to increased storm runoff many sewage systems cannot cope with the resulting peak flow without investment in greater capacity.

Deforestation is also a cause of increased flood runoff and a decrease in channel capacity. This occurs due to an increase in deposition

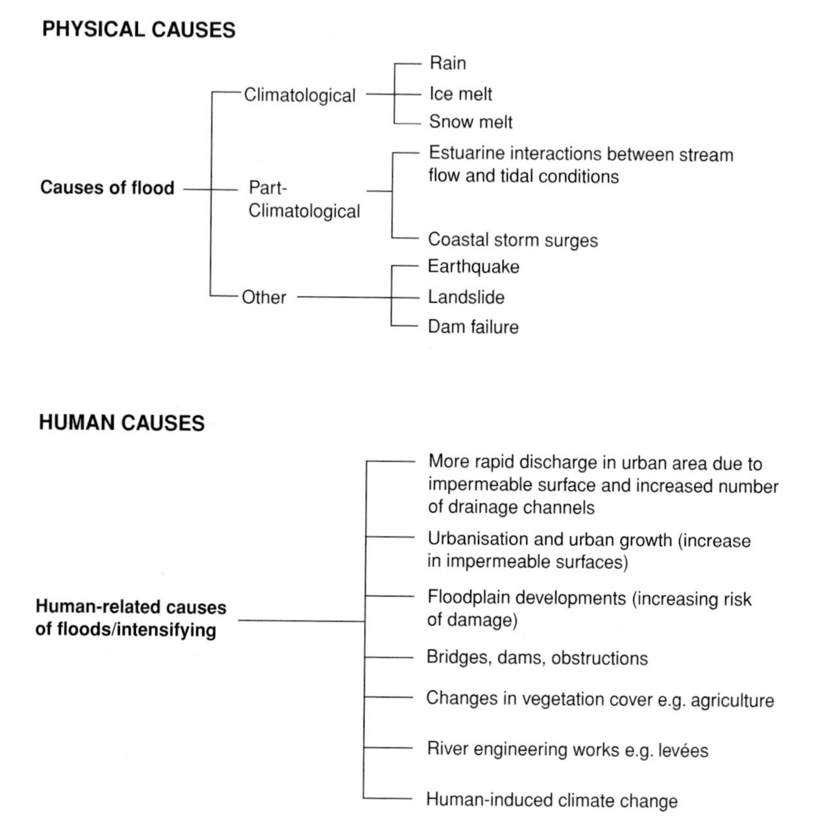

Figure 28 Causes of floods

within the channel. However, there is little evidence to support any direct relationship between deforestation in the Himalayas and changes in flooding and increased deposition of silt in parts of the lower Ganges–Brahmaputra. This is due to the combination of high monsoon rains in the Himalayas, steep slopes and the seismically unstable terrain, which ensure that runoff is rapid and sedimentation is high irrespective of the vegetation cover.

4 Human Causes of Floods

Economic growth and population movements throughout the 20th century have caused many floodplains to be built on. However, in order for people to live on floodplains there needs to be flood protection. This can take many forms such as loss-sharing adjustments and event modifications.

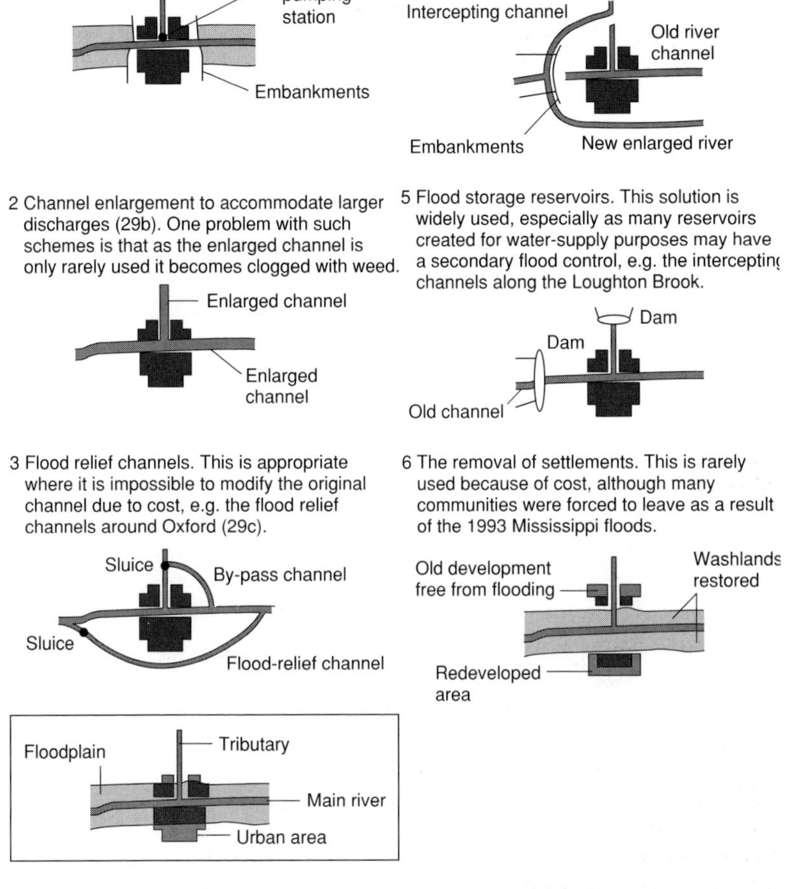

Figure 29a Flood diversion channels

Loss-sharing adjustments include disaster aid and insurance. Disaster aid refers to any aid, such as money, equipment, staff and technical assistance, that are given to a community following a disaster. However, there are many taxpayers who argue that they cannot be expected to fund losses that should have been insured.

In developed countries insurance is an important loss-sharing strategy. However, not all flood-prone households have insurance and many of those are insured may be underinsured. In the floods of central England in 1998 many of the affected households were not insured against losses from flooding because the residents did not

Figure 29b Levees and reinforced steel banks

Figure 29c Flood diversion channel at Osney, Oxford

believe that they lived in an area that was likely to flood. Hence, they had very limited flood insurance.

Event-modification adjustments include environmental control and hazard-resistant design. Physical control of floods depends on two measures – flood abatement and flood diversion. Flood abatement involves decreasing the amount of runoff, thereby reducing the flood peak in a drainage basin. This can be achieved by weather modification and/or watershed treatment, for example, to reduce flood peak over a drainage basin. There are a number of strategies including:

- reforestation
- reseeding of sparsely vegetated areas to increase evaporative losses
- mechanical land treatment of slopes such as contour ploughing or terracing to reduce the runoff coefficient
- comprehensive protection of vegetation from wild fires, overgrazing, clear cutting of forests, land or any other practices likely to increase flood discharge and sediment load
- clearance of sediment and other debris from head water streams
- construction of small water and sediment holding areas
- preservation of natural water detention zones.

Flood diversion measures, by contrast, include the construction of levees, reservoirs and the modification of river channels (Figure 29c). Levees are the most common form of river engineering. They can also be used to divert and restrict water to low value land on the floodplain. Over 4500 km of the Mississippi River has levees and channel improvements such as enlargement to increase the carrying capacity of the river. Reservoirs store any excess rainwater in the upper drainage basin. Large dams are expensive and may well be causing earthquakes and siltation. It has been estimated that some 66 billion m^3 of storage will be need to make any significant impact on major floods in Bangladesh!

5 Hazard-resistant Design

Flood-proofing includes any adjustments to buildings and their contents which help to reduce losses. Some are temporary such as blocking up of certain entrances, use of shields to seal doors and windows, removal of damageable goods to higher levels and the use of sandbags (Figure 30). By contrast, long-term measures include moving the living spaces above the likely level of the floodplain. This normally means building above the flood level, but could also include building homes on stilts (Figure 31).

Bewdley on the River Severn has been flooded 100 times in the past century, the 2000 crisis being the worst since 1947, and so engineers are testing a new portable steel fence here to tackle flooding. The Environment Agency is also testing a £3 million flood defence system of steel planks that can be slotted into vertical supports when the river begins to swell. The supports are bolted on to unobtrusive steel base plates that are the only visible sign of flood defences. The plates are set into the pavement on top of an underground waterproof flood wall built within the quayside. The barriers have proved very successful in Europe, but they are new in Britain.

Figure 30 Sandbags in action

Figure 31 Caribbean police station on stilts

6 Forecasting and Warning

During the 1970s and 1980s flood forecasting and warning had become more accurate, and are now one of the most widely used measures to reduce the problems caused by flooding. Despite advances in weather satellites and the use of radar for forecasting, over 50% of all of unprotected dwellings in England and Wales have less than 6 hours of flood warning time. In developed countries flood warnings and forecasts reduce economic losses by as much as 40%. In most developing countries, however, there is much less effective flood forecasting. An exception is Bangladesh. Most floods in Bangladesh start in the Himalayas, so authorities have about 72 hours warning.

CASE STUDY: FLOODS IN BANGLADESH

Much of Bangladesh has been formed by deposition from three main rivers – the Brahmaputra, the Ganges and the Meghna. The sediment from these and over 50 other rivers forms one of the largest deltas in the world, and up to 50% of the country is located on the delta (Figure 32). As a result much of the country is less than 1 metre above sea level and is under threat from flooding and rising sea levels. To make matters worse, Bangladesh is a very densely populated country (over 900 people per km^2) and is experiencing rapid population growth (nearly 2.7% per annum). It contains nearly twice as many people as the UK, but is only about half its size. Hence, the flat, low-lying delta is vital as a place to live as well as a place to grow food.

Almost all of Bangladesh's rivers have their source outside the country. For example, the drainage basin of the Ganges and Brahmaputra cover 1.75 million km^2 and includes parts of the Himalayas and Tibetan Plateau and much of northern India. Total rainfall within the Brahmaputra–Ganges–Meghna catchment is very high and very seasonal – 75% of annual rainfall occurs in the monsoon between June and September. Cherrapunji, high in the Himalayas, has an average annual rainfall of over 10,000 mm, and this can rise as high as 20,000 mm in a 'wet' year. Moreover, the Ganges and Brahmaputra carry snowmelt waters from the Himalayas. This normally reaches the delta in June and July. Peak discharges of the rivers are immense – up to 100,000 m^3/s (i.e. 100,000 cumecs) in the Brahmaputra, for example. In addition to water, the rivers carry vast quantities of sediment. This is deposited annually to form temporary islands and sand banks.

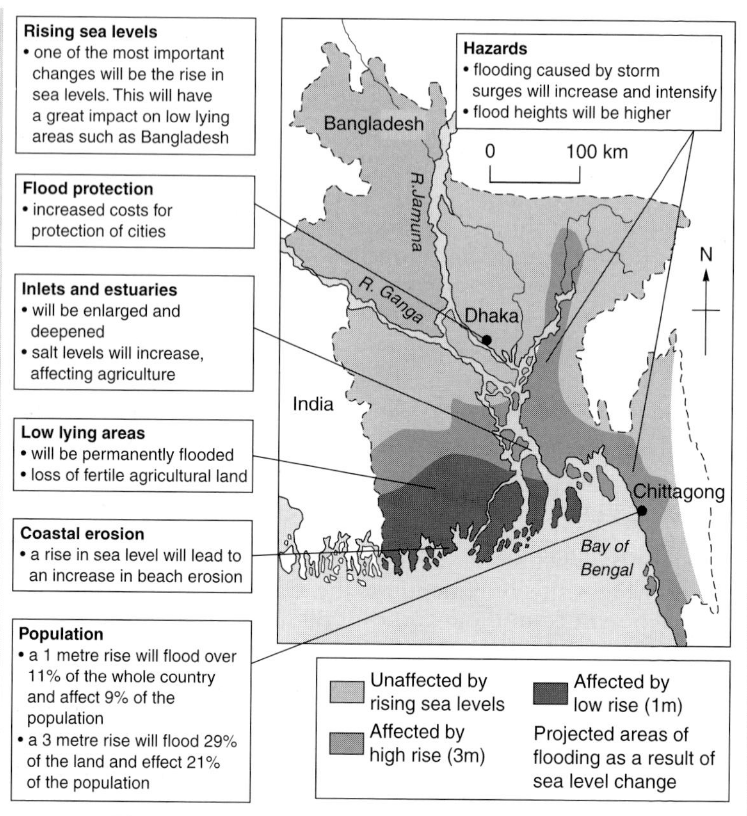

Figure 32 Map of Bangladesh to show flood hazard

The advantages of flooding

During the monsoon between 30% and 50% of the entire country is flooded. The flood waters:

- replenish groundwater reserves
- provide nutrient-rich sediment for agriculture in the dry season
- provide fish – fish supply 75% of dietary protein and over 10% of annual export earnings
- reduces the need for artificial fertilisers
- flush pollutants and pathogens away from domestic areas.

The causes

There are five main causes of flooding in Bangladesh – river floods, overland runoff, flashfloods, 'back-flooding' and storm surges. Snowmelt in the Himalayas combined with heavy monsoonal rain causes peak discharges in all the major rivers during June and July. This leads to flooding and destruction of agricultural land. Outside the monsoon season, heavy rainfall causes extensive flooding which may be advantageous to agricultural production, since it is a source of new nutrients. In addition, the effects of flash floods, caused by heavy rainfall in northern India, have been intensified by the destruction of forest which reduces interception, decreases water retention and increases the rate of surface runoff. Human activity in Bangladesh has also increased the problem. Attempts to reduce flooding by building embankments and dykes have prevented the back flow of flood water into the river. This leads to a ponding of water (also known as 'drainage congestion') and back-flooding. In this way, embankments have sometimes led to an increase in deposition in drainage channels, and this can cause large-scale deep flooding. Bangladesh is also subject to coastal flooding. Storm surges caused by intense low-pressure systems are funnelled up the Bay of Bengal.

The Flood Action Plan

It is impossible to prevent flooding in Bangladesh. The Flood Action Plan attempts to minimise the damage of flooding, and maximise the benefits of flooding. The Plan relies upon huge embankments that run along the length of the main rivers. At an estimated cost of $10 billion they could take 100 years to build. Up to 8000 km of levees are planned for the 16,000 km of river in Bangladesh. However, they are not able to withstand the most severe floods, for example those in 1987 and 1988, but provide some control of flooding. The embankments contain sluices that can be opened to reduce river flow and to control the damage caused by flooding.

The embankments are set back from the rivers. This protects them from the erosive power of the river, and has the added advantage of being cheap both to install and maintain. In addition, the area between the river and embankment can be used for cereal production.

Nevertheless, the Flood Action Plan is not without its critics. There are a number of negative impacts of the scheme. These include:

- increased time of flooding, since embankments prevent backflow into the river
- not enough sluices have been built to control the levels of the flood waters in the rivers – this means that there may be

increased damage by flooding if the embankments are breached, since the rapid nature of the breach is more harmful than gradual flooding
- sudden breaches of the embankments may also deposit deep layers of infertile sand, thereby reducing soil fertility
- compartmentalisation may reduce the flushing effect of the flood waters, increasing the concentration of pollutants from domestic effluents and agrochemicals
- by preventing back-flow to the river, areas of stagnant water will be created which may increase the likelihood of diseases such as cholera and malaria
- embankments may cause some wetlands to dry out, leading to a loss of biodiversity
- decreased flooding will reduce the input of fish, which is a major source of protein, especially among the poor.

The rivers of Bangladesh are, in part, controlled by factors beyond the country. There is a delicate balance between the disadvantages that the rivers create, such as death and destruction, advantages that the rivers bestow, such as a basis for agriculture and export earnings. To date there has been little agreement as to how to control the peak discharges of the rivers. The Flood Action Plan uses embankments to control the distribution and speed of flooding, although the embankments have, in turn, led to serious social, economic and environmental problems.

7 Land-use Planning

Most land-use zoning and land-use planning has come in the last 30–40 years. Land-use management has been effective in protecting new housing developments in the USA from losses up to the one in 100 year flood (i.e. the floods that we would expect to occur more than once every century). In England and Wales floodplain development has been controlled by the Town and Country Planning Acts since 1947. In Britain there hasn't been the same encroachment onto the floodplain as there has been in the United States. Partly, this is due to population growth, for example between 1952 and 1982 the population in England and Wales grew by 12% compared to 50% in the USA, 73% in Canada and 78% in Australia, therefore in Britain less new building has been required, whereas in the United States and Australia the opposite is true and hence more encroachment onto floodplains. Nevertheless, there are increasing worries in the UK concerning the frequency of winter flooding and floodplain development.

One example where partial urban relocation has occurred is at Soldier's Grove on the Kickapoo River in south-western Wisconsin, USA. The town experienced a series of floods in the 1970s, and the Army Corps of Engineers proposed to build two levees and to move part of the urban area. Following floods in 1978 they decided that relocation of the entire business district would be better than just flood damage reduction. Although levees would have protected the village from most floods, they would not have provided other opportunities. Relocation allowed energy conservation and an increase in commercial activity in the area.

CASE STUDY: PROTECTING THE MISSISSIPPI

It is not just the delta that receives attention from engineers, but the whole of the Mississippi. For over a century the Mississippi has been mapped, protected and regulated. The river flows through 10 states and drains one-third of the USA. It contains some of the USA's most important agricultural regions. A number of methods have been used to control flooding in the river, and its effects, including

- stone and earthen levees to raise the banks of the river
- holding dams to hold back water in times of flood
- lateral dykes to divert water away from the river
- straightening of the channel to remove water speedily.

Altogether over $10 billion has been spent on controlling the Mississippi, and annual maintenance costs are nearly $200 million. But it was not enough. In 1993, following heavy rain between April and July, many of the levees collapsed allowing the river to flood its floodplain. The damage was estimated to be over $12 billion yet only 43 people died. Over 25,000 km^2 of land were flooded. The river was only performing its natural function – people and engineers had modified the channel so much that its normal function did not occur often, and so people thought they were safe from the effects of flooding.

Managing the Mississippi Delta

Deltas are difficult areas to manage. The example of the Mississippi (Figure 33) illustrates this very clearly. The delta covers an area of 26,000 km^2, drains an area of 3,220,000 km^2, and carries 520 km^3 of water and 450 million tonnes of sediment into the Gulf of Mexico every year. Much of the 210 million tonnes of sediment is carried in suspension. About 40% of the load is silt and 50% clay. The silt is deposited to form a bar up to 7 m high, whereas the clay is transported further out the delta. Deposition takes place during floods forming levees and floodplains, and is enhanced by biological activities, notably colonisation by plants and the trapping of sediment by vegetation. Sediment-charged waters break through the natural levees (crevasse splaying) to find shorter steeper routes, a process known as avulsion, the channel then splits (bifurcates) and sediment is deposited, and hence, the delta grows, e.g. distributaries in West Bay grew 16 km between 1839 and 1875.

Deltas follow a cycle of development. This may take anything between 100 and 1000 years. First, new channels are created as a result of crevasse splays. Sediment is deposited and new land is created. As a result of changes in gradient, channels are abandoned, deposition declines and a new area is developed. However, this pattern is disrupted by human activities. According to some geographers, if the Mississippi were left to its own devices a new channel would have been created by the mid-1970s, so much so that the ports at New Orleans and Baton Rogue would be defunct. However, river protection schemes have prevented

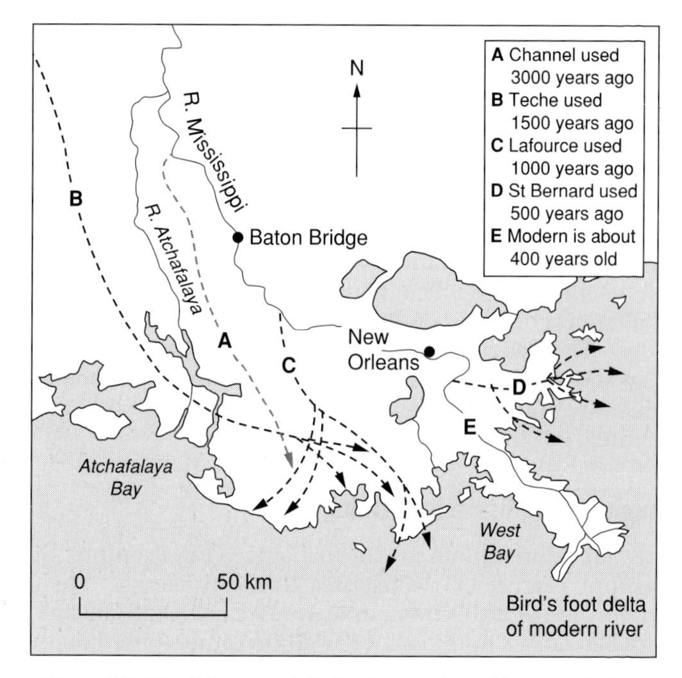

Figure 33 The Mississippi Delta (reproduced by permission of Hodder Arnold)

this. For example, at New Orleans 7 m levees flank the river and 3 m levees abut Lake Pontchartrain. New Orleans is 1.5 m below the average river level and 5.5 m below flood level!

The Mississippi Delta is retreating at rates of up to 25 million tonnes/year. It accounts for 40% of US wetlands and over 100 km^2 of wetlands are being lost each year. The cause of the delta's decline is a mixture of natural and man-made conditions: rising sea levels, subsidence (due to the weight of the delta on the earth's crust), groundwater abstraction, tropical storms and changes in the location of deposition are all causing the delta to change. Increasingly, it appears that the delta is trying to abandon its current course and to develop a new course along the Atchafalaya Channel.

Flood relief measures include:

- Bonnet Carre Waterway from New Orleans to Lake Pontchartrain and then to the Gulf of Mexico
- Atchafalaya River which carries up to 25% of the Mississippi flow, 135 million tonnes of sediment (but this is a fertile area and famed for its bayou landscape and culture)
- Dredging of the Atchafalaya and Mississippi Rivers
- Morganza floodway between Mississippi and Atchafalaya.

8 Multipurpose Water Development – The Case of Large Dams

Much of what we have discussed so far relates to the removal or deflection of flood water away from vulnerable locations. Another form of management is that of water storage and regulation. This has the advantage of containing the water in such a way that it can be used for many purposes other than flood control, for example transport, irrigation, hydroelectric power and industrial development. Of the more than 39,000 large dams (exceeding 15 m) in existence, almost 90% have been built since 1950. They offer development benefits through hydroelectric power, drinking water supplies, flood control, and recreational opportunities, however, some of the impacts are unforeseen, as shown in Figure 34(a and b). Some reservoirs are silting up rapidly as a result of soil erosion upstream and increased deposition behind the reservoir (Figure 35). The effects of compulsory relocation due to the construction of large-scale infrastructure projects have generally been negative.

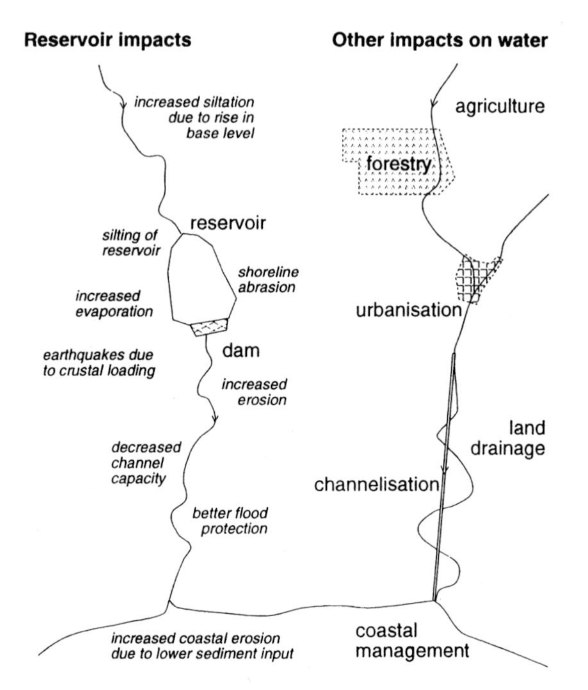

Figure 34a The possible impacts of dam construction on the environment

Dam	River and country	Impacts
Akosombo	Volta, Ghana	Bilharzia, river blindness, 70,000 dispossessed
Aswan	Nile River, Egypt	Bilharzia, siltation, erosion, salinisation of the soil
Kainji	Niger River, Nigeria	44,000 displaced, fishing effects; loss of biodiversity
Kariba	Zambezi River, Zambia/Zimbabwe	Flooding of lands, 50,000 people dispossessed salvinia weed infestation
Katse	Malibamatso River, Lesotho	25 households displaced, 2700 people lost arable land, 80% lost grazing land and a variety of natural resources
Kiambere	Tana River, Kenya	6000 dispossessed, reduced incomes, livestock, land loss and loss of wildlife, including rare primates
Maguga	Komati River, Swaziland	66 households displaced, loss of arable land grazing and valuable wood, river sand and other natural resources
Manatali	Senegal River, Mali	Health impacts such as malaria, social tensions interethnic conflict increased

Figure 34b African dam projects and their impacts

Logistics	Benefits	Side-effects	General assessment and prospects
• Land development was a necessity to cope with the incredible imbalance between the country's population growth and agricultural production. Agriculture is basic to the Egyptian economy.	• Controlled high floods and supplemented low ones; saved Egypt from the monetary cost to cover the damage from both high and low floods.	• Water loss through evaporation and seepage is likely to affect the water supply needed for development plans; studies show that the water loss is within the predicted volume.	• The Aswan High Dam is solid engineering work; more importantly it is fulfilling a vital need for 40 m people.
• Egypt found no option but to increase the water supply for land development policies of both horizontal and vertical extension,	• Allowed for increased in cultivated area through reclamation and increased the crop production of the existing land through conversion from basin to perennial irrigation.	• Loss of the Nile silt would require costly use of fertilisers; it has also caused river bed degradation and coastal erosion of the northern delta.	• All dams have problems; some are recognised while others are unforeseen at the time of planning.
• Building a dam and forming a water reservoir in an Egyptian territory minimises the risks of water control politics on the part of riparian countries.	• Improved Nile navigation and changed it from seasonal to year-round.	• Soil salinity is increasing and land in most areas is becoming waterlogged, due to delays in implementing drainage schemes.	• The lessons learned from the Aswan project, however, is that dams may be built with missionary zeal but little careful planning and monitoring of side effects.

(Cont'd)

Figure 35 The Aswan Dam – a summary sheet

Logistics	Benefits	Side-effects	General assessment and prospects
• Egypt perceived the Aswan High Dam as a multipurpose project, basic to national development plans.	• Electric power generated by the dam now supplies 50% of Egypt's consumption, but the dam was built primarily not for power generation but for water conservation	• Increased contact with water through irrigation extension schemes is expected to affect the Schistosomiasis rates adversely; evidence exists that rates have not increased due to use of protected water supplies.	• As a result of the new semi-capitalist policies and also the technical and monetary aid that Egypt received from several Western countries, it is expected that several of the dam's problems will be efficiently controlled and monitored.
	• The new lake resources are potentially economic, including land cultivation and settlement, fishing and tourist industries.	• The Nile water quality has been deteriorating; studies indicate the occurrence of change in the water quality parameters; they do not constitute a health hazard at present.	• Dam-related studies have given recognition to present problems and have provided possible solutions. The dam would meet its expectations on several aspects providing that research findings are utilised. • The development potential and economic returns of this water project are expected to be very rewarding in the long run, if the project's developments are systematically studied and monitored.

Figure 35 *(Cont'd)*

CASE STUDY: WATER RESOURCE DEVELOPMENT IN TURKEY – THE CASE OF ILISU

Having seen some of the advantages and disadvantages of earlier schemes such as the Nile, new schemes have been subjected to much greater scrutiny. A very good example of this is the Ilisu Dam in Turkey (Figure 36). Ilisu is part of the South-east Anatolia Project (GAP), a giant hydropower and irrigation scheme on the Euphrates and Tigris rivers in the Kurdish part of Turkey. The potential for conflict concerning Turkey's proposed water development is very great. There is a particular problem with Syria and Iraq over plans to develop the Euphrates River. Turkey is developing more dams, reservoirs and irrigation schemes thereby reducing the flow of the Tigris and the Euphrates. Syria and Turkey have quarrelled over the Euphrates since the 1960s, and both Syria and Iraq have protested over Turkey's water development plans. Syria claims that Turkey interrupts water flow, causing power cuts and threatening agriculture. The potential for conflict is enormous and likely to escalate. But it is not just with international problems that Turkey has made the headlines; many of its projects cause social,

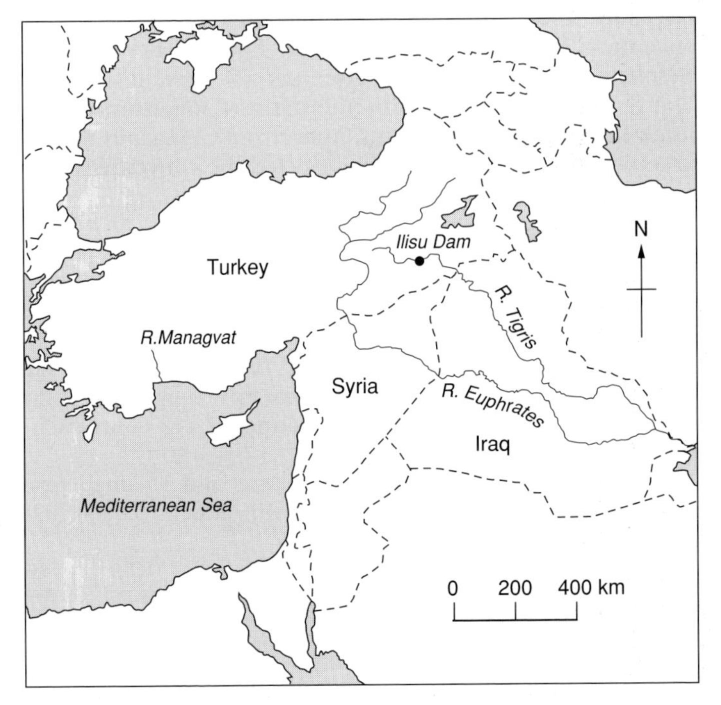

Figure 36 The Ilisu Dam

economic and other problems within Turkey itself. The Ilisu Dam project, for example, is a controversial one for a variety of political, social, environmental, economic and archaeological reasons.

The site of the Ilisu Dam is located on the Tigris River 65 km upstream of the Syrian and Iraqi border. Ilisu is currently the largest hydropower project of Turkey. A rockfill dam 1820 m long and 135 m high will create a reservoir with a surface area of 313 km^2. The Ilisu power station could have a capacity of 1200 MW. The costs are estimated to be $1.52 billion. Construction was due to start in mid-1999, and production of power in mid-2006.

Political problems

The claims of Turkey, Syria and Iraq on the Euphrates and Tigris exceed the capacities of the two rivers by 55% and 12%, respectively. The dams on Euphrates are used primarily for irrigation, and they reduce the average annual water flow by almost 50%. In contrast, the Tigris projects, primarily used for power production, reduce water flows by 10%. Turkey has not been prepared to compromise regarding the management of the rivers. Instead it uses its location on the upstream part of the river to pressurise and blackmail the other riparian countries.

In May 1997, the UN General Assembly approved the Convention on the Non-Navigational Uses of Transboundary Waterways. This convention attempts to prevent significant negative impacts of projects on international waterways on other countries in the drainage basins. Apart from China and Burundi, Turkey was the only country that rejected the convention

Social problems

GAP reservoirs such as Ataturk or Karakaya have so far involuntarily displaced 100,000s of people. Compensation has usually been tied to the property of land or houses. Since most land in South-east Anatolia is concentrated in the hands of large landowners, many landless families were not compensated at all. Instead, they quietly moved to the slums of big cities such as Diyarbakir or Istanbul.

The Ilisu reservoir will flood 52 villages and 15 small towns, including the city of Hasankeyf, and will affect 15,000–20,000 people.

Environmental problems

Solid waste and wastewater of major cities such as Diyarbakir (population of 1 million), Batman and Siirt are being dumped into the Tigris without any treatment.

The Ilisu reservoir will also infest the area with malaria and leishmaniosis. The annual sediment load of the Tigris is 15–30 million m^3 and this would fill up 10–20% of the reservoir's normal operating capacity within 50 years. The authors of an Environment Impact Assessment (EIA) expect the useful life of the reservoir to be 80–100 years. Empirically, reservoir planners have often underestimated the sedimentation rate.

Archaeological impacts

If and when it is completed, the Ilisu reservoir will flood Hasankeyf, a Kurdish town with a population of 5500. Hasankeyf is the only town in Anatolia that has survived since the Middle Ages without destruction. It contains relics of Assyrian, Christian, Abyssinian-Islamic and Osmanian history in Turkey.

Lacking analysis of alternatives

At a cost of $1300/kW (plus financing costs), Ilisu is a relatively expensive power project. Project opponents in Turkey believe that power could be saved at a lower cost by modernising the country's transmission system, which has a reputation of being inefficient. According to the authors of the EIA, no supply-side or demand-side alternatives to Ilisu have been considered as part of the feasibility studies. It seems likely that the Turkish government is prepared to pay a high price for Ilisu because of its interests to control the Kurdish population of South-East Anatolia, and to increase its political clout *vis-à-vis* Syria and Iraq.

The Ilisu project appears to violate five World Bank guidelines. The guidelines in question are the Environmental Assessment, Environmental Policy for Dam and Reservoir Projects, Involuntary Resettlement, Projects on International Waterways and Management of Cultural Property.

In November 2001 the UK construction firm Balfour Beatty (the leader of the international consortium that was to develop the Ilisu Dam) pulled out of the £1.25 billion project. The firm announced that it was withdrawing from the project because the 'commercial, social and environmental issues were unlikely to be resolved soon'. It is believed that the project had the personal backing of the British Prime Minster, partly as reward for its help in providing bases for the bombing of Afghanistan earlier in 2001. Following the World Commission on Dams report in July 2001 Skanska, the Swedish member of the consortium withdrew. The withdrawal of funds suggests that the scheme has little prospect of being followed through. However, the Turkish government has always insisted that it would continue with the project whether or not it had foreign backing.

CASE STUDY: ENGINEERING SOLUTIONS: THE COLORADO

Some of the world's largest rivers do not reach the sea. In the wake of economic development of the communities along rivers comes an increase in water consumption that depletes the rivers of their reserves. At the extreme end of the spectrum, the Amu Darya and Syr Darya (see page 15) in Central Asia that feed the Aral Sea – have been deprived of close to their entire water reserves for cotton irrigation. Another good example is the Colorado (Figure 37).

The 2333 km river drains an area of over 630,000 km^2. It was the first river in the USA to be used for multipurpose development. A number of large dams, such as the Hoover and Parker dams, were built in the early 20th century. State or federal authorities fund the dams, no holistic approach has been adopted. The dams are used for:

- irrigation – 800,000 hectares of farmland are now fed by the Colorado
- flood and silt control – the Colorado once carried 130 million tons of suspended sediment to the sea, now it hardly carries anything. Flood peaks and discharge have been reduced since the dams have been built.
- power – the dams produce 120 million kW of electricity
- domestic and industrial water supplies – the Colorado now supplies the demands of over 40 million people in seven US states and Mexico
- recreation – the lakes behind the dams attract many tourists.

The Colorado is one of the first rivers in the world to have its entire flow used up before it reaches its estuary in the Gulf of California.

Distance from source (km)	River flow (m^3/s) 1920	River flow (m^3/s) 2000
100	5	200
300	124	220
600	409	370
1200	905	430
2000	1003	170

Figure 37 Changing water flow in the Colorado

Management problems

There have been many problems associated with the management of the Colorado.

- Water supply has failed to meet demand. Up to 25% of water is lost through evaporation and leakage, and more through

Controlled flooding of the Grand Canyons has prompted planners to flush out other American river valleys where dams have created environmental problems. The results of the Grand Canyon experiment so far are promising. Years of sluggish flow in the Colorado River had allowed silt to accumulate in the wrong places, and altered the habitats downstream.

When the Glen Canyon Dam was completed in 1963, it ended the spring floods in the Grand Canyon. The dam operators release water at times of peak demand for electricity, with flow varying during the day and reaching a maximum in summer. But the pattern of water flow is now slow and unnatural. Where the river's periodic floods used to dump sand along the river's banks, forming beaches, now the sand just sinks to the bottom.

These concerns prompted the government to start the Glen Canyon Environmental Studies project in 1982. A decade of research showed that periodic floods are necessary to maintain the vitality of riverside ecosystems. A consensus emerged that the best way to restore river beaches and ecosystems was by releasing water through the dam in 'controlled floods'.

Planners selected the seven days between March and April, the time of natural spring floods. The dam can release only 1300 cubic metres per second, about half the average peak flow of natural spring floods, sufficient to redistribute sand from the river bottom and restore beaches, sand bars and critical habitats. The water was allowed to flow at full flood for several days and was then slowed for the final two days.

The canyon's ecosystem has been rejuvenated as a result of these floods. The flood created 55 new beaches, and added to three-quarters of the existing ones. The biggest gains were in the first 100 kilometres below the dam. The flood washed away the vegetation that had grown along the riverbanks during years of low water.

The flood also helped to rejuvenate marshes and backwaters, which are critical habitats for native fish and some endangered species such as the south-western willow flycatcher and the humpbacked chub. Nevertheless, marshes can take years to become established.

One of the goals of the flood was to try to rid the river of alien species of fish such as striped bass and carp, but the torrent was not strong enough to flush them out.

Plans to restore other American rivers include the Columbia River, Snake and Missouri Rivers.

Figure 38 Controlled flooding of the Grand Canyon

wasteful irrigation techniques. Significant increases in population in the west and south-west of America have put further pressure on supplies. The Central Arizona Project, for example, now distributes 1.85 trillion litres of water each year to farms, Indian reservations, industries and residential homes over a distance of 570 km.

- Sediment has become trapped behind the dams, which is expected to shorten their lives.
- Once-established wildlife is disappearing along stretches of the river. The Colorado Delta, for example, is now dry for much of the year and has become starved of sediment, as a result bird, plant and fish populations which previously thrived are now dying out.
- Recycling methods have resulted in some farmers receiving highly saline water that has been damaging to their crops. Desalinisation techniques are expensive.
- Flooding has occurred, but it has been largely attributed to human mismanagement of the dams.

Nevertheless, it has been possible to reduce some of the problems caused by the dams. For example, controlled flooding of the Grand Canyon (Figure 38) has allowed the partial rehabilitation of the Grand Canyon's ecosystem.

Summary

This chapter has shown the causes of floods and examined some of the attempts to manage them.

- Floods are caused by a variety of environmental factors such as heavy rain, snow melt and coastal surges.
- Human factors intensify the impacts of floods. Deforestation, over-cultivation and urbanisation, for example, all lead to greater flood peaks.
- There are a number of strategies for flood alleviation. These depend on level of economic development as well as the personal characteristics of the decision-maker.
- The main options are adjustment, abatement (in the catchment) and protection.
- Traditionally, hard engineering structures such as dams, channel modification (realignment, widening deepening), levees and wing dykes have been used.
- These may have unforeseen impacts such as decreased water quality, increased sedimentation, bed and bank erosion, and loss of habitats
- The impact of dams is economic, social, environmental and political.

Achieving consensus on international rivers is not easy. A number of guiding principles have been established

- equity of distribution among riparian countries
- equity means fair shares which depend on a number of factors such as size of basin; climate conditions; economic and social needs of the users; population growth; comparative costs of developing alternative forms of water in member states; the avoidance on undue waste and damage.
- Increasingly, people are aware of the negative impacts of large engineering structures and in some cases there are attempts to limit the damage to the environment.

Questions

1. Comment on the geographical pattern of fatalities, observed losses and insured losses for hazards in 2002, as shown in Figure 26.

Figure 34(a) shows some of the possible impacts of dam construction on the environment.

2. Describe and explain any **two** effects on the hydrological cycle.
3. Suggest what is meant by the term 'clear water erosion'.
4. Explain why clear water erosion increases below the dam.
5. Give reasons to explain why the construction of large dams may lead to shoreline abrasion (coastal erosion) and earthquake triggering.

5 Soft Engineering

KEY DEFINITIONS

Abstraction Removal of water from surface water or groundwater, usually by pumping.
Baseflow The flow in a river derived from emergent groundwater and spring discharges.
Biodiversity A rich variety of living plants and animals.
Blue-green algae Organisms with some properties characteristic of both bacteria and algae – natural inhabitants of many inland waters.
Buffer zone A strip of land adjacent to a river that is free from any development.
Catchment The total area of land that drains to a specified watercourse or waterbody.
Environmental indicator A measure that can be used to assess the present state of the environment by looking at trends over time.
Floodplain encroachment Development on low-lying land adjacent to a river where water is naturally stored during flood conditions.
Groundwater Water which is contained within the pores and crevices of soils and rocks.
Riparian Of or on the bank of a river; or relating to the legal rights of the landowner of a riverbank.
Restoration A complete structural and functional return to a river's natural state
Rehabilitation A partial structural and functional return to the river's natural state
Sustainable development Management of the environment to meet the needs of the present without compromising the ability of future generations to do the same.

1 Introduction

In the last chapter we saw how hard engineering structures have had many unforeseen impacts. In this chapter we look at alternative methods of river control – first, the management of whole catchments (catchment management plans) and, second, at soft engineering projects – plans that work with nature rather than against it. Nevertheless, it is not always possible to classify schemes as either hard engineering or soft engineering since in some catchment management plans there may be elements of both.

2 The Riparian Zone

Riparian land is any land that adjoins or directly influences a body of water. It includes:

- the land immediately alongside small streams and rivers, including the riverbank
- areas surrounding lakes
- wetlands and river floodplains that interact with the river in times of flood.

Depending on the nature of the land (floodplain, gorge or valley) and the adjacent land use (national park, farming, forestry, urban housing), the width of riparian land that needs special management will range from very narrow to a wide, densely vegetated corridor.

There are a number of reasons why maintaining a healthy riparian zone is important. These include decreased erosion, improved water quality, healthy ecosystems, maintaining river courses, stock management, decrease in insect pests, increase in capital value, opportunities for diversification, retention of nutrients, lowered water tables, increased fish stocks, landscape refuge, decreased algal growth and ecotourism

Riparian zones act as corridors of natural vegetation, preventing species becoming isolated and dying out. Riparian vegetation also helps to reduce light and temperature levels of stream ecosystems. This controls the growth of algae, even when nutrient levels in the stream water have increased.

Riparian vegetation can absorb and use natural or added nutrients that might otherwise be washed into streams. Also, deep-rooted riparian vegetation, such as willow, may lower water tables along riverbanks, reducing the movement of salt and nutrients into streams from subsurface flows.

The roots of vegetation reinforce the soil. Fine roots are more important in this process than thick roots. Tree roots can substantially increase soil strength to a depth of at least 2 metres, and to a distance equivalent to the circumference of the canopy (the dripline).

Riparian vegetation also uses much of the water present in streambanks, and improves the drainage of streambank soil (Figure 39). Often banks collapse when they are saturated with water, so that riparian vegetation, by using water and improving drainage, helps reduce the risk of sudden collapse. The speed of water flow in a channel can be decreased by vegetation growing either on the bank or in the water, and by debris or sediment in the stream. However, if the vegetation becomes too dense, for example with reed beds, it can direct flow into the banks and increase scour. In large channels the key is to establish vegetation as far down the face of the bank as possible as well as on the top of the bank. This reduces the amount of erosion of the bank.

Feature of vegetation	Impact on amount of bank collapse	Impact on amount of scour
Root systems	Large reduction	Large reduction
Water use	Small reduction	N/A
Buttressing	Significant reduction	N/A
Reduced flow speed	N/A	Large reduction
Surcharge	May be reduced or increased	N/A

Figure 39 Impact of vegetation on bank collapse and scour

a) Improving water quality

A number of materials from riparian lands can contaminate streams. This includes soil particles, nutrients, plant material and microbes. In rural areas, soil erosion is the most important cause of reduced water quality.

Sediment and certain nutrients, notably phosphorus, are transferred to rivers by overland flow. Dissolved nutrients, such as nitrogen, can also move through the soil as base flow.

The clearing of catchments during forestry operations or urban development has led to substantial increases in the amounts of sediment entering streams and rivers. This may contaminate water supplies, cover breeding sites for fish and, by filling up stream pools, deprive animals of the deep waters that are a vital refuge in dry seasons and during droughts.

b) Riparian vegetation and water quality

Vegetation within the riparian zone can slow the overland movement of water, and cause sediment to be deposited on the land before it reaches the river channel. Riparian vegetation can also take up and remove some of the nutrients being transported. Deep vegetation use significant quantities of subsurface waters. Hence, riparian vegetation can influence underground water flows and the nutrients entering the stream by this route.

Riparian vegetation also provides shade to streams and therefore regulates stream temperature. Natural vegetation and grassy strips can trap around 90% of the sediment moving from upslope, and they are equally effective in trapping or absorbing nutrients

Factors affecting the amount and type of sediment moving in overland flow include soil type, intensity of land use, presence of livestock, vehicle tracks or gullies which generate sediment, and the likelihood of the overland flow being concentrated into a narrow pathway. Sound land management to maintain good water quality includes maintaining adequate vegetation cover, use of contour farming and maintenance of general soil cover elsewhere within the catchment, and the timing and placement of fertilisers to avoid periods of intense runoff. Uncontrolled use of riparian lands by stock contributes significantly to water quality. If not managed carefully, cattle will spend long periods along streambanks, leading to overgrazing and baring of the soil surface. Direct inputs of nutrients from cattle through manure and urine add substantially to the loads of nitrogen and phosphorus within the stream. These nutrients can then support excessive growth of weeds and algae.

c) Shading

Where channels are less than about 10 metres wide, they may be partly or completely shaded by adjoining riparian vegetation. Decreased light intensity reduces excessive growth of weeds and algae in river systems. The dim or patchy lighting resulting from riparian vegetation also assists in providing habitats within the stream.

Once natural riparian vegetation is cleared for agricultural or urban development, light intensity within the river increases greatly. This disturbs the natural ecosystem, and can lead to excessive growth of algae or massive growth of water plants. Excessive in-stream growth eventually decomposes and this leads to a decrease in oxygen, fish kills, and further loss of native species and reduction in water quality.

d) Temperature

As well as reducing light intensity, shading maintains lower water temperatures, reducing the daily maximum by as much as 10°C. This is important, as many indigenous plants and animals are sensitive to wide fluctuations in temperature and are unable to grow or reproduce when temperatures rise above a critical point.

e) Riparian habitats for wildlife

Riparian lands differ from adjacent areas in many ways. They often have better soils, higher moisture and different plant species. Riparian land often contains a high biodiversity of living organisms, and plays a crucial role as a corridor for the movement of plants and animals. Although riparian lands may occupy only a small percentage of the catchment landscape, they are vital to its long-term health and sustainable land management.

Moisture is an important habitat feature of riparian lands, which generally occupy the lower parts of the landscape where there is usually more soil moisture available for plant growth, and it is retained for longer periods. The soils are also rich in organic matter, with a better supply of nutrients available to support plant growth. Soils and nutrients may be replenished by periodic flooding. As a result of the greater availability of water, and the presence of soils that are rich in nutrients, riparian lands are amongst the most productive ecosystems on earth.

However, because riparian lands are highly productive parts of the landscape, they have been extensively cleared for agricultural development. They can also be subject to intense pressure when used for grazing by domestic stock and native and feral animals. Uncontrolled access by domestic stock to riparian land can cause increased runoff, bank erosion, loss of productive land, decline in important wildlife habitat, reduced water quality and damage to in-stream ecosystems. Riparian lands have been modified extensively by human use for urban development and recreation. Complete or over-clearing is a common problem and, where native riparian vegetation does exist it is often in poor condition.

f) Pollution in rivers

Blue-green algae (cyanobacteria) are a natural part of river systems. When in balance they are not a problem, but increased nutrients and low flows have contributed to severe algal bloom outbreaks in many river systems. The conditions that favour algal blooms are well understood. Algal growth depends on the availability and supply of the nutrients, nitrogen (N) and phosphorus (P), light and warm water temperatures. These cause blooms of blue-green algae often coinciding with long periods of warm, sunny weather, high nutrient levels and still water. The enrichment of waters with nutrients is known as eutrophication (see Figure 40).

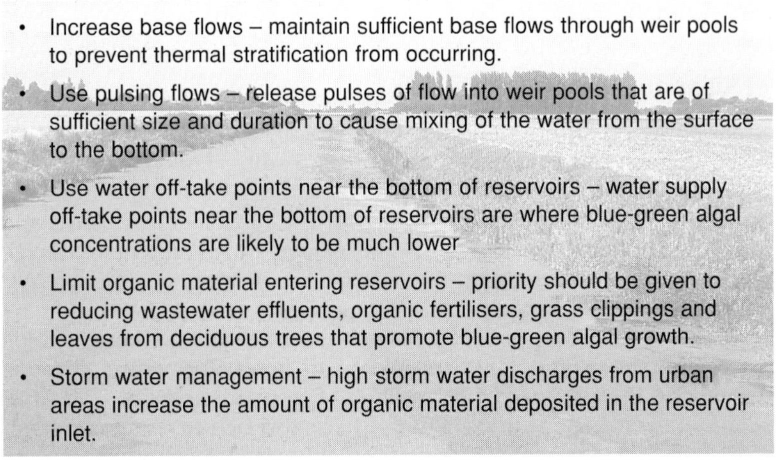

- Increase base flows – maintain sufficient base flows through weir pools to prevent thermal stratification from occurring.
- Use pulsing flows – release pulses of flow into weir pools that are of sufficient size and duration to cause mixing of the water from the surface to the bottom.
- Use water off-take points near the bottom of reservoirs – water supply off-take points near the bottom of reservoirs are where blue-green algal concentrations are likely to be much lower
- Limit organic material entering reservoirs – priority should be given to reducing wastewater effluents, organic fertilisers, grass clippings and leaves from deciduous trees that promote blue-green algal growth.
- Storm water management – high storm water discharges from urban areas increase the amount of organic material deposited in the reservoir inlet.

Figure 40 Management practices to combat eutrophication

Rivers rarely experience algal bloom outbreaks during periods of high flow. This means that new approaches to manipulating the flow of rivers and water reservoirs may hold the key to preventing algal blooms, saving millions in water treatment costs and environmental damage caused by algal bloom outbreaks.

During storms and flood events, large amounts of nutrients (N and P) are carried to rivers from surface and subsurface erosion of soils. These nutrients are either recycled within the water, or released from the sediments. Algal growth is sustained by nutrients and, once a large flood event is over, the combination of increased nutrients and low flows create ideal conditions for algal bloom outbreaks to occur.

g) Phosphorus in catchments

Phosphorus is an essential component of all plants and animals, and is a natural part of the rocks that comprise the Earth's crust. Changes in land use (e.g. intensive agricultural development) have increased the amounts of phosphorus transferred to rivers, reservoirs and lakes. There is a close relationship between how land is managed (Figure 41) and the impact phosphorus may have on in-stream health. Phosphorus enters rivers from a number of different sources.

Management practices developed to minimise or intercept erosion (whether surface or subsurface) are also likely to minimise phosphorus transport. The following management practices can help reduce the generation and delivery of phosphorus in catchments.

- Focus on controlling diffuse sources of phosphorus such as streambank and gully erosion
- Stabilise streambanks and control stock access to reduce the risk of bank collapse.
- Develop engineering structures such as contour banks, gully sediment traps, artificial wetlands and dams to reduce on-site erosion and sediment transport
- Manage erosion in high flow events to control the transport of phosphorus to downstream river reaches and other receiving waters.
- Develop practical methods to reduce flood peaks and volumes
- Pasture management to limit the transfer of phosphorus-rich soils to streams.
- Manage grazing (e.g. stocking rate and intensity) to limit erosion of soils due to stock tracks, groundcover destruction and excess surface erosion.
- Provide stock watering and shade away from drainage lines to limit destabilisation and erosion of streambanks.

Figure 41 Management practices to control phosphorus transfer to streams

Sources of phosphorus include:

- 'point' sources such as sewage treatment plants, intensive animal industries, and irrigation and storm water drains
- 'diffuse' sources such as soil and fertiliser runoff.

As a general rule, point sources are the major contributor of phosphorus to waterways in urban environments, whereas diffuse sources dominate in rural environments.

CASE STUDY: CATCHMENT MANAGEMENT IN AUSTRALIA

a) Changes to rivers

Following European colonisation, floodplains and catchments were farmed, creating a number of important changes to flows and the water regime:

- forests were cleared for agriculture, increasing runoff and raising water tables, leading to dryland salinity and greater quantities of salt entering rivers
- levees were built to protect farms from flooding, but have proved ineffective against large floods. Levees have isolated the river from its floodplain
- wetlands were drained to allow farming thereby destroying habitats
- weirs, dams and floodplain banks were built to catch and divert water for agricultural use
- the practice of capturing water in winter and releasing it in summer for irrigation has overturned the normal flow pattern in southern regulated rivers
- river regulation has produced a much more constant flow in summer as water is transmitted to irrigation areas
- excessive water extraction for irrigation has severely disadvantaged downstream users
- the construction of block banks or bunds across extensive areas of floodplain in Queensland and New South Wales to capture water for irrigating cotton has significantly reduced floodplain wetting downstream.

b) Streambank stability

Streambank erosion is a natural process as streams meander across the landscape. Since European settlement of Australia, however, the rate of streambank erosion has increased markedly. There are two main reasons for this:

- extensive clearing of deep-rooted, natural vegetation from catchments for agricultural and urban development – this has led to increased overland runoff, so stream channels can no longer contain flood peaks, thus bank and bed erosion occurs
- widespread removal of riparian vegetation from streambanks, either through deliberate clearing for development, or through the combined effects of stock grazing and fire – the removal of large, woody debris – have made streambanks unstable.

These have weakened the ability of streambanks to resist the erosive forces of increased flood flows and resulted in eroding streambanks becoming a common feature in Australian landscapes.

Streambank erosion often involves the loss of valuable agricultural and recreational land. As the banks collapse or are eroded and washed away, sediment and nutrient loads increase and water quality is reduced.

c) Managing woody debris in rivers

There is increasing evidence that, before European settlement, most rivers in Australia had a large amount of large woody debris (LWD) along their banks and within their channels. To early settlers, LWD was a nuisance. It made access to streams by stock difficult, and it was a major hazard to transport and navigation. It was generally believed that LWD blocked the channel and caused additional flooding at times of peak flow. As a result, particularly in southern Australia, much LWD was removed from streams and burnt.

Research over the past 20 years has shown that woody debris is a vital component for the healthy functioning of rivers. LWD also helps to protect the beds and banks of streams from erosion and in many situations it does not contribute significantly to flooding. For example, a large trunk which extends across all, or most of the channel, will cause an upstream pool to develop preventing further erosion during major flow events. The large quantities of LWD present in many river channels in their natural state would have been sufficient to armour and protect the bed against erosion.

LWD also has many important ecological benefits. LWD helps to trap leaf litter and other organic matter moving down the stream to form a major source of food for animals. Water flowing over LWD becomes aerated, and the range of flow rates produced around debris is important for the diversity of plant and animal life required for healthy rivers. Large debris is also vital for the survival and growth of many important fish species. It provides habitat and shelter from predators, while hollow logs are an essential spawning habitat for several native fish species; for example the Mary River Cod of south-east Queensland, and the River Blackfish of Victoria and Tasmania require submerged hollow logs in which to lay and nurse their eggs.

LWD does lead to some erosion and reshaping of channels, especially through the formation of scour pools immediately adjacent to the obstruction. The presence of LWD can sometimes increase or decrease local bank erosion. The size and orientation of the debris, velocity and depth of flow, and the character of the material making up the streambed and bank, all influence the potential for erosion.

d) Inland rivers

Australia's inland rivers flow across the vast, mainly arid regions of the continent. Some, like the Murray-Darling flow out to sea, whilst others like the Diamantina and Cooper Creek, flow inland to Lake Eyre. Inland rivers share many characteristics (Figure 42).

For example, during dry periods, the Paroo River in Australia is limited to a series of waterholes and some permanent freshwater lakes; however, large floods may see the river inundate nearly 800,000 hectares (Figure 43).

Australia's inland rivers differ from most of the rest of the world's rivers because of their high flow variability. This is largely due to the highly unpredictable climate of Australia's interior. Cooper Creek and the Diamantina River are among the world's most variable major rivers. They have 'boom' (floods) and 'bust' cycles (dry periods), and everything in between. Inland rivers

Inland rivers are

- are highly variable with boom or bust cycles
- function at different spatial and temporal scales
- have high biodiversity, especially on the floodplains
- are mainly used for pastoral and rangeland grazing enterprises
- are in areas of low population density with scattered towns and settlements
- are under threat by water resource development through the building of dams, weirs and levees for irrigation
- have had little research investment
- cross state government boundaries and present particular problems for management
- have mainly low flow quantity

Figure 42 Characteristics of inland rivers (adapted from *Land and Water Australia*, 2002)

	Narran Lakes	Paroo River
Environmental and cultural values	Narran Lakes is a wetland of international significance and is well known for its birdlife. The area is especially important for indigenous populations, and grazers depend on the floods for their livelihoods.	The wetlands of the Paroo River are important for their birdlife, with Currawinya National Park of international importance. The area is especially important for indigenous populations, and grazers depend on the floods for their livelihoods.
Water resource development and long-term environmental health	In the late 1980s there were only a small number of off-river storage schemes upstream of the Narran Lakes. By 1993 the storage volume had increased to 300,000 Ml, by 1996 it was 600,000 Ml and by 1999 950,000 Ml. Water diversions have reduced the flow to the lake by 74%. This is reducing and/or degrading the area of floodplain and habitats by about 50%, and an associated loss of biodiversity.	Very little water is diverted from the Paroo River. Irrigation licenses proposed in the mid-1990s met considerable opposition. Subsequent legislation has tended to protect the river.
Legislation and policy	Levees, pumps and storage schemes were used to regulate the flow of water but without proper environmental assessment, or strong legislation. Flood waters were not controlled by government-built structures so the waters were free for anyone to use, once they had developed some form of scheme.	Although many of legislation and policies that allowed development of the water upstream of the Narran Lakes were present on the Paroo River, better understanding of development impacts and public pressure resulted in a water management plan for the Paroo that had an ecological focus and a agreement between the states to protect the river's flow.
Communication	There was effectively very limited understanding on the community's regarding legislation, government policy, long-term impacts of the scheme. By contrast, the farming industry of Queensland was politically motivated and exerted significant pressure in favour of irrigation schemes.	Key to communications was a workshop involving floodplain grazers, conservation organisations and scientists. By contrast, the irrigation lobby was poorly represented in the catchment and therefore exerted little pressure. The combined force of the conservationists, local community and scientists was able to help formulate a policy which helped protect the river rather than allow its abuse.

Figure 43 The Narran and the Paroo

have long periods of low or no flow, followed by periods of extreme flooding.

Some inland rivers end in wetlands, such as Lake Eyre, Gwydir wetlands and Lake Gregory, while others flood billabongs (pools) and forests such as the Barmah-Millewa forest and the Macquarie Marshes. Similarly, inland rivers' floodplains can be subdivided

- freshwater lakes, such as Coongie Lakes and the Paroo River's overflow lakes
- saline lakes such as Lake Eyre
- large floodplain swamps such as the Macquarie Marshes and the Barmah-Millewa Forest.

People have relied on inland rivers for thousands of years and their reliance is still as important as ever (Figures 44 and 45). European colonists used them for navigation, irrigation and urban development. Most rural towns in Australia are built on reliable waterholes such as Birdsville on the Diamantina River and Bourke on the Darling River. Most of the water is used for farming, although increasingly more is used for industrial uses and for recreational uses such as swimming.

There are many threats to inland rivers. The most serious includes water regulation via dams and weirs, as well as abstraction, and the construction of levees. In addition, the introduction of exotic species, such as the European carp, increased salinity levels, sedimentation from catchment development, increased levels of pollution, increased grazing pressures, and removal of logs and branches out of the rivers (desnagging) are all capable of changing the nature of inland streams and rivers.

Most dams are built in upland areas and so reduce the amount of water that makes it to lowland wetland floodplains. About half of the flow of the Murray River is extracted and this has adversely affected the ecology of the river in its lower course. The Murray-Darling Basin covers approximately 14% of Australia, and is Australia's most developed river basin. A total of 80% of the total flow from rivers in the Basin is diverted. Dams were built early in the 20th century on the southern rivers of Murray and Murrumbidgee to help farmers. Irrigation uses 70% of all water used in Australia and 95% of all water in the Murray-Darling Basin. However, most of the large dams were built after the 1950s and in the northern part of the Basin where rainfall is higher. These dams include those on the Macquarie-Bogan, Namoi, Gwydir, Border Rivers and Lachlan. The last dam to be built in the Basin was the enlargement of the Pindari Dam on the Borders River in 1995.

A set of principles for inland river management may assist the community, policy makers and managers to make decisions for a sustainable future. Principles upon which to base inland river management and policies could include recognition that

1. naturally variable flow regimes, the dry phase, and the maintenance of water quality are fundamental to the health of inland river ecosystems;
2. flooding is essential to ecosystem processes and makes a significant contribution to pastoral activities;
3. structures such as dams, weirs and levees can have a significant impact on the connectivity along rivers and between the river and its floodplain; solutions are needed to either minimise these impacts or find alternatives;
4. water is essential to rural industries and communities, who have the responsibility at the local level to manage water resources;
5. catchment management, and integrated surface and groundwater management, are important concepts that need to be put into practice;
6. sufficient knowledge exists to ensure that water resource allocation decisions are made on a sustainable basis, and a strong commitment is needed to access and fully utilise best available scientific information;
7. new developments should be undertaken only after appraisal indicates they are economically viable and ecologically sustainable; promoting greater water efficiency is essential to achieving sustainable industries;
8. high conservation value rivers and floodplains need to be identified, and is some cases protected in an unregulated state;
9. stressed rivers need to be identified, and priorities established for their rehabilitation;
10. improved institutional and legal frameworks are needed to meet community river management aspirations;
11. with all parties making a commitment to work together, management regimes can be developed that are ecologically, economically, socially and culturally sustainable.

Figure 44 Principles for inland river management

e) Pollution in Australian rivers

In a national study of 24 rivers in Australia, the links between river flow and blue-green algae abundance were researched and two dominant trends emerged:

- as flows decreased blue-green algae abundance increased – this was clearly shown in the temperate rivers of NSW and Victoria
- in tropical rivers in Queensland, prolonged low flow conditions led to more blue-green algae being present.

West of the old gold-rush city of Bendigo towards the Murray River lies a scene of slow strangulation by salt. Dead trees stick out of rivers, salt pans form on roadsides and the brick walls in old buildings turn white as salt rises up out of the ground.

The local water tables appears to be rising remorselessly and salt comes to the surface with them. They are destroying some of Australia's most productive farmland and its biggest river system, and even threaten to deprive Adelaide, the capital of South Australia, of fresh water. For as far as the eye can see, the hills and plains are bare. Settlers cut down the trees for use in mineshafts when the state of Victoria was gripped by gold fever in the 1850s. Farmers moved in with sheep, cattle, crops and irrigation schemes. The result, 150 years later, is a looming shortage of drinkable water.

Australia is the world's driest continent, and water has always figured deeply in the national psyche. Ever since settlers started moving to the interior, Australians have dreamed of turning coastal rivers inland, but only one scheme has ever been carried out. In a gargantuan piece of engineering, the flow of the Snowy River in south-eastern New South Wales has been reversed, so that instead of its water being lost to the Pacific, it now runs through tunnels dug under mountains, is then used for power generation and ultimately goes to irrigate dry inland plains. In the 50 years since it was undertaken, the Snowy Mountains scheme has been regarded as a proud symbol of Australia's development, and of the pioneering spirit of its people in turning hostile elements to their advantage. Only now are the true costs emerging.

A recent report by the Murray-Darling Basin Commission found that salt levels were rising at a more alarming rate than anyone had realised, thanks to tree-clearing, farming and irrigation.

The cost is put at almost A$1 billion ($638m) a year in damaged land and lost production; and in 20 years, says the commission, many of the basin's rivers may be unusable even for irrigation, let alone as a source of drinking water. The land too is suffering. Now that so much salt is being pushed to the surface, it is simply staying there, rather than being flushed out to sea.

Finding a solution is proving to be a political nightmare. The Snowy River provides an example. Since its diversion, the Snowy's run has been reduced to a trickle, in some parts to just 1% of its original flow. The solution may be a radical one: turning intensively irrigated land over to crops such as olives and grapes, which need less water. At present, however, few politicians seem prepared to tell farmers of boom crops such as cotton and rice in the heavily irrigated Murray-Darling basin that their time is up. Nature may soon do it for them.

Figure 45 Unsuitable farming, deforestation and too much irrigation threatens parts of Australia with ecological devastation

There were no cases where blue-green algae were recorded during high flow periods.

During low flows, stratification develops and the water forms layers with a warm surface layer on top of a cold bottom layer. Stratification often develops during the summer months due to high solar radiation during the day. In addition, many of Australia's inland rivers have high concentrations of suspended clay. This reduces the amount of light reaching the lower parts, thereby limiting algal growth to the surface. Once the blue-green algae float in the well-lit surface they cut off the light from the layers below. This eventually leads to the decline of the ecosystem since light is an important source of energy for all ecosystems.

f) Phosphorus

Much of the diffuse phosphorus in Australian catchments is due to soil erosion. In contrast, only where there are high population densities and intensive agriculture is there strong evidence that phosphorus comes directly from sewage, fertilisers and animal wastes. The amount of phosphorus available depend upon:

- the geology of the catchment
- the overlaying soils and their natural phosphorus concentrations
- land-use type and intensity
- the nature and magnitude of the erosion process.

Some rocks, such as basalts, have naturally high amounts of phosphorus. The weathering and break-down of basalt results in natural phosphorus present in the overlying soils. When this soil is cultivated, increased gully erosion can occur, delivering large amounts of phosphorus into river systems.

Phosphorus moves through the landscape either in solution, or in suspension attached to soil particles (particularly fine clays). Overland runoff is especially important in areas with high rainfall intensities (such as the tropics in northern Australia) and where soils are intensively worked. Over 85% of the sediment-bound phosphorus in far north Queensland comes from hill slope erosion of surface soils. By contrast, approximately 50% of the sediment-bound phosphorus in the Murray-Darling Basin and other catchments in New South Wales and Victoria are derived from a combination of gully and channel erosion. In parts of the Murray-Darling Basin, where gullies are widespread, most of the soil-bound phosphorus is washed from the gullies during large storms.

Catchments with a high drainage density and high natural phosphorus concentrations will generate high sediment yields and total phosphorus. A significant amount of the phosphorus will be derived from erosion within the channels. In contrast, catchments with low drainage density and low natural phosphorus will have low overall sediment and phosphorus entering rivers. Under these conditions, a greater proportion of this material will be delivered from surface erosion.

3 Managing Rivers: The Role of Catchment Management Plans in the UK

Catchment Management Plans (CMPs) are plans for the integrated planning, management and development of the water environment. The Environment Agency has a nationwide programme for the completion of the first round of Catchment Management Plan (CMPs) by the end of 1998. The CMP identifies appropriate levels for growth in the river catchment. CMPs can only be achieved through a partnership between a number of key agencies and organisations, such as town and country planners, local authorities, environmental groups and landowners. The funding involves a variety of sources. The first practical application by the NRA was for the Cotswold Water Park in the Upper Thames CMP.

The EA has a duty to conserve and enhance the environment when carrying out any of its functions, and a further duty to promote conservation and enhancement more widely. For water resources, this includes overall policies on water resources, water quality and surface water management, and flood management. Catchment Management Plans have already achieved a number of objectives:

- an outline of future plans for each catchment allowing it to reach its full environmental potential
- the identification of conflicts such as between flood defence systems and preserving river ecosystems
- bringing together different interest groups, such as the private water companies.

Catchment Management Plans vary significantly. Each is site specific but all share a common purpose of trying to control poor land uses and to encourage better practises. Some plans have been developed for small catchments (such as the Wandle below), whereas others refer to much larger catchments (such as the Thames). Each plan identifies the main issues that need to be tackled as well as suggesting how they should be tackled.

a) Agenda 21

The United Nations Conference on Environment and Development (1992) resulted in Agenda 21 schemes. Agenda 21 are the plans drawn up by local councils to achieve sustainable developments. The links between Agenda 21 and Catchment Management Plans are strong. Agenda 21 allows:

- increased emphasis on environmental considerations when assessing planning applications
- increased community involvement in development issues
- the development of Catchment Management Plans

Where development would lead to a risk of water resources further allocations of land should normally be resisted until adequate resources can be made available.

Current Sustainability principles for water resources
- There should be no long term deterioration of the water environment resulting from water use or water resources for future use
- Reasonable demands for water from both existing and new social and economic development should be satisfied
- Priority should be given to the management of water demand, and to ensure the best use is being made of existing resources. Only if additional water resources are still required will new water resource development be considered
- In managing water resources, opportunities to enhance the water environment should be identified

Guidance for development plans
Where new development would be at direct or unacceptable risk from flooding or would aggravate the risk of flooding elsewhere to an unacceptable level proposals should be resisted

Current sustainability principles for flood defence
- Effective defence for people and property against flooding from rivers and the sea should be provided, together with adequate arrangements for flood forecasting and warning
- Inappropriate development within floodplains should be resisted where such developments would be at risk from flooding or may cause flooding elsewhere
- Flood defence is an intervention in natural processes and therefore a balance has to be struck between maintaining and supporting natural floodplains and alleviating flood risk.
- Floodplains should be safeguarded to protect their vital role in allowing for the storage and free-flow of flood waters
- To minimise and increased surface water run-off, new development must be carefully located and designed. Where appropriate, source control measures should be incorporated into the scheme.

Figure 46 Guidance for development plans

- the opportunity for integrated catchment management planning
- protection of the environment
- increased public awareness of the water environment
- help to identify appropriate environmental indicators.

The Environment Agency manages surface water management, mineral extractions, sewage capacity, water efficiency and demand, waste disposal, habitat enhancement and restoration, and research and development. For example, in the Thames Region there is a need for more new housing, for development of derelict sites, for mineral extraction, flood defences, a supply of safe drinking water and sustained agricultural yields.

The National Rivers Authority (NRA), and subsequently the Environment Agency (EA) developed the idea of Catchment Management Planning. This requires a Catchment Management Plan (CMP) for each natural river catchment throughout England and Wales. The plan entailed data evaluation, issues analysis, external liaison and consultation, and the CMP provides a vehicle to focus attention on the water environment. By their very nature CMPs involve the close contact with local communities and other interested organisations.

CASE STUDY: MANAGING AN URBAN RIVER: THE WANDLE VALLEY

Any attempt to manage the River Wandle must therefore balance the conflicting needs of the many interests there are. Flood storage, pollution control, conservation and recreation must be developed alongside the need to dispose of waste materials, and the need to find more mineral resources for Britain's expanding housing market.

The Wandle has a number of assets:

- it provides a valuable green corridor through an urban area
- it links a number of sites of conservation interest such as Beddington Sewage Treatment Works, Wilderness Island Nature Reserve and Watermeads
- much of it has been described as a Site of Metropolitan Interest due to the variety of its plant life
- water quality is classified by the EA as poor to fair
- there is widespread contamination of groundwater, largely as a result of past industrial activity, e.g at Beddington.

Key Catchment Planning Issues

There are a number of sites for redevelopment along the River Wandle providing opportunities for environmental enhancements and access to the river frontage.

The proposed gravel extraction and landfill operations will have a large impact, but opportunities should be identified for enhancing the River Wandle corridor. In particular, the proposals present an opportunity to create a large body of water for recreation in an area of London identified by the Sports Council as deficient in this respect. Informal access could also be improved.

The proposal for waste disposal is of concern both in terms of pollution and flooding. The guidelines set out in the NRA's 'Policy and practise for the protection of groundwater' must be followed to try to protect the groundwater and surface water in the area from pollution. Any further encroachments into the floodplain should be resisted in order to retain the capacity, as far as possible, and allow the free flow of the flood waters. Increases in surface water run-off caused by development should be contained by using source control measures so that the risk of flooding is not increased further.

The diversity of flora and fauna should be protected and enhanced where development provides the opportunity.

(Source: NRA, Thames 21, 1995)

Wandle management

The vision of the Wandle Catchment Management Plan is to achieve and maintain an improved state of well-being for the Wandle, Beverley Brook and Hogsmill river catchments by working in partnership with all interested parties to resolve increasingly conflicting demands on water uses (Figure 47). Such a vision can only be realised through active community participation.

There are a range of pressures resulting from intense urbanisation in terms of floodplain encroachment, water usage, effluent disposal and demand for water-related recreation must be balanced against the need to protect and enhance the diverse ecology, industrial/archaeological heritage and landscape associated with the local water environment.

The key strategic objectives are to:

- maintain and improve surface and groundwater quality throughout the catchment
- protect low flows from further reduction and investigate alleviation
- integrate the management of environmentally sensitive flood defence works with the control of surface water runoff
- protect and enhance permanent water bodies throughout the catchment
- improve management techniques for in-stream and bankside riverine habitats

Population		985,000 (approx)
Catchment area		339 km^2
Urban/suburban area		132 km^2 (39%)
Length of river (source to R.Thames)	(Wandle)	19.0 km
	(Beverley Brook)	14.3 km
	(Hogsmill)	9.9 km
Water resources		
Average annual rainfall		694 mm
Average flow	(River Wandle)	143 Ml/d
	(Beverley Brook)	47 Ml/d
	(Hogsmill River)	84 Ml/d
Total licensed groundwater abstraction		240 Ml/d (estimate)
Water quality		
River length	Class A	0 km
	Class B	1.8 km (3%)
	Class C	14.1 km (32%)
	Class D	14.5 km (33%)
	Class E	11.7 km (26%)
	Class F	2.4 km (6%)
Flood defence		
Length of statutory main river	(Wandle)	26.7 km
	(Beverley Brook)	23.1 km
	(Hogsmill)	9.5 km
Area at risk from flooding once every 50 years (observed and predicted)		5.0 km^2

Figure 47 Catchment characteristics

- seek to minimise the environmental impact of water control structures
- balance the needs between the recreation and conservation uses of rivers and stillwaters
- maintain and improve the fisheries status of rivers within the catchment
- promote community pride in the local water environment.

The Wandle, Beverley Brook and Hogsmill river catchments cover an area of 339 km^2 between the River Thames and the dip slope of the North Downs to the south. Home to nearly a million people the catchment is predominantly urban in character with housing development continuing to exert pressure on the water environment (Figure 48). The decline of heavy industry has paid environmental dividends in terms of reducing industrial effluent

Figure 48 The urban Wandle

disposal, but treated sewage effluent disposal from such a highly concentrated population has inevitably had an impact on surface water quality.

During low flow conditions treated sewage effluent can account for in excess of 90% of the river flow in the Wandle and Beverley Brook. The impermeable nature of urban areas combined with floodplain encroachment has also resulted in a fundamental alteration to the natural hydrological regime with river levels rising very quickly during storm events.

Groundwater abstraction has affected baseflows of the Wandle and Hogsmill rivers, which are both spring-fed. River water from the Wandle at Goat Bridge is used to top-up flows in the Carshalton ponds under low flow conditions. The natural quality of the chalk groundwater is still high, but localised groundwater within the gravel aquifers to the north of the catchment has been heavily polluted by industry. The future disposal of solid waste to backfill pits excavated for gravel in the Mitcham and Beddington area is under review.

The combination of poor water quality caused by urban storm water runoff and treated sewage effluent, increased stream velocity and engineered concrete river channels has impacted heavily on the ecological status of many reaches of river particularly with regard to fisheries' status. A large proportion of the watercourse length is only of poor or very poor biological quality. The primary cause of the low biological quality is the poor chemical water quality due to discharges from the sewage treatment works. Surface water runoff from the urban areas and frequent pollution incidents also contribute to the poor biological scores. Some chemical pollution incidents may have long-term effects on biological quality.

Figure 49 Semi-rural environment along the Wandle, London

The Worcester Park STW discharge also causes an occasional midge problem in the river within Kingston. The suspended solids discharged from the STW constitute a rich and abundant food source for midge larvae which reach exceptionally high densities. Jet washing of the river bed every 2–3 months is effective at artificially reducing the density in the short term. Regular algal blooms interfere with the amenity value of the Richmond Park Ponds. Between 1990 and 1993, for example, there were 45 significant pollution incidents in the catchment

Making the most of the recreation, amenity and education opportunities offered by the water environment within such a heavily populated catchment is a prime objective, e.g. the micro-turbine on the River Wandle as a modern-day example of the historical use of the river for water power. Many parts of the Wandle retain a semi-natural environment (Figure 49), an important asset for recreational and leisure activities.

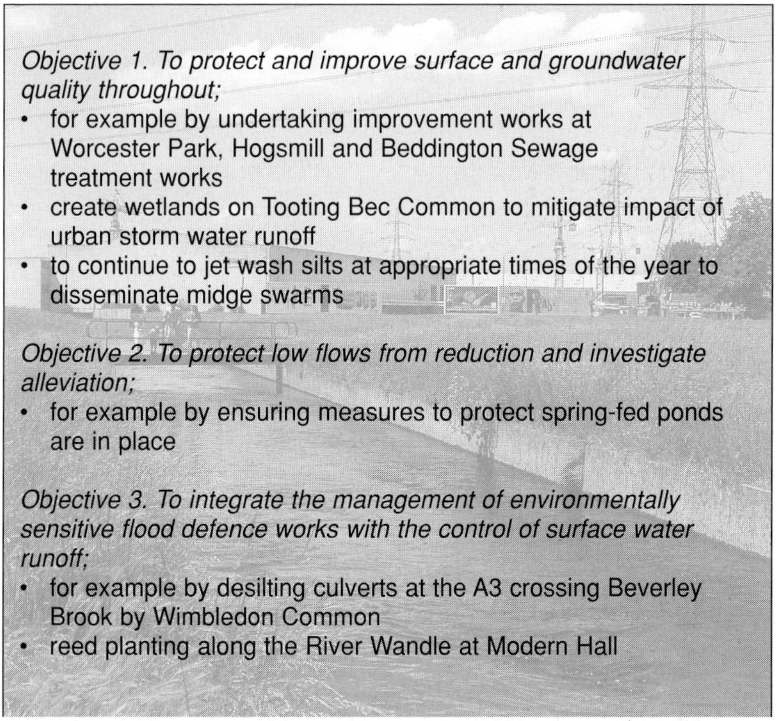

Objective 1. To protect and improve surface and groundwater quality throughout;

- for example by undertaking improvement works at Worcester Park, Hogsmill and Beddington Sewage treatment works
- create wetlands on Tooting Bec Common to mitigate impact of urban storm water runoff
- to continue to jet wash silts at appropriate times of the year to disseminate midge swarms

Objective 2. To protect low flows from reduction and investigate alleviation;

- for example by ensuring measures to protect spring-fed ponds are in place

Objective 3. To integrate the management of environmentally sensitive flood defence works with the control of surface water runoff;

- for example by desilting culverts at the A3 crossing Beverley Brook by Wimbledon Common
- reed planting along the River Wandle at Modern Hall

Figure 50 Objectives and examples

Objective 4. To protect and enhance permanent water bodies;

- for example Environment Agency, Local Authorities and all interested parties to resolve conflict between users of Wimbledon Park Lake and other water bodies

Objective 5. To improve management techniques for instream and bankside habitats and landscapes;

- update River Corridor Surveys for River Wandle, Beverley Brook and Hogsmill River
- eradication of invasive Japanese knotweed from Hogsmill River

Objective 6. To maximise the environmental benefit of water control structures;

- promotion of policies for consideration in development site planning briefs to aid sustainable development of Wandle tidal creek area

Objective 7. To balance the needs between the recreation and conservation uses of controlled waters;

- Environment Agency to work with all interested parties in identifying suitable buffer zones for river reaches

Objective 8. To maintain and improve fisheries status;

- to continue a five year rolling programme of surveys on the River Wandle tidal creek, Beverley Brook, Hogsmill, and River Wandle

Objective 9. To promote community pride in the local water environment.

- Promote Environment Agency 'Riverwork' primary school teaching pack, 'Sources' secondary school teaching pack and National River Watch schemes
- revise and update factfiles on Wandle, Beverley Brook and Hogsmill catchment.

Figure 50 *(Cont'd)*

4 River Restoration

It is far more cost-effective to keep streams and rivers clean than allowing them to deteriorate and then have to undergo expensive rehabilitation and restoration. Indeed, it is clear that clean rivers still require management to keep them healthy, and to prevent them from deteriorating. The principle of 'protect first, restore second' is a good one. The first rule of rehabilitation is to avoid the damage in the first place. It is expensive, difficult and a very slow process restoring rivers, whereas it is relatively cheap, easy and quick to keep them healthy.

The term restoration implies that the river is returned to its original quality. Defining river quality is not easy for it includes a wide range of attributes such as water chemistry, sediment and flow regime, plants and animals present, and the health of neighbouring riparian areas. Restoration may not be possible in many cases (since rivers had a pristine quality before human activities affected them) so it may only be possible to rehabilitate rivers instead.

Rehabilitation refers to an improvement in the river quality, although it may not reach the same quality as the original river. It is effectively a pragmatic approach to improving river standards, i.e. there is recognition that it is impossible to reach the original standard but it is possible to improve on the current condition of the river. This may mean fixing (improving) only certain aspects of the stream. In contrast, remediation recognises that the river has changed so much that the original condition is no longer relevant and a new condition is designed (Figure 51). Remediation aims to improve the ecology of the stream, but the end result may not necessarily resemble the original stream. In practice, most restoration schemes will only partially restore or rehabilitate the river due to the large number of human-related uses in the floodplains (buildings, farmland and transport, for example).

River restoration schemes are becoming increasingly common as the benefits of natural rivers and their floodplains are realised. The aims of the River Restoration Project (RRP) are

- to establish international demonstration projects which show how the state-of-the-art restoration techniques can be used to recreate natural ecosystems in damaged river corridors
- to improve understanding of the effects of restoration work on nature conservation value, water quality, visual amenity recreation and public perception
- to encourage others to restore streams and rivers.

The RRP is an independent organisation backed by scientific and technical advisers drawn mainly from organisations connected with rivers and river environments. Its aims are to restore and enhance damaged rivers for conservation, recreation and economic use, returning them as closely as possible to their natural condition.

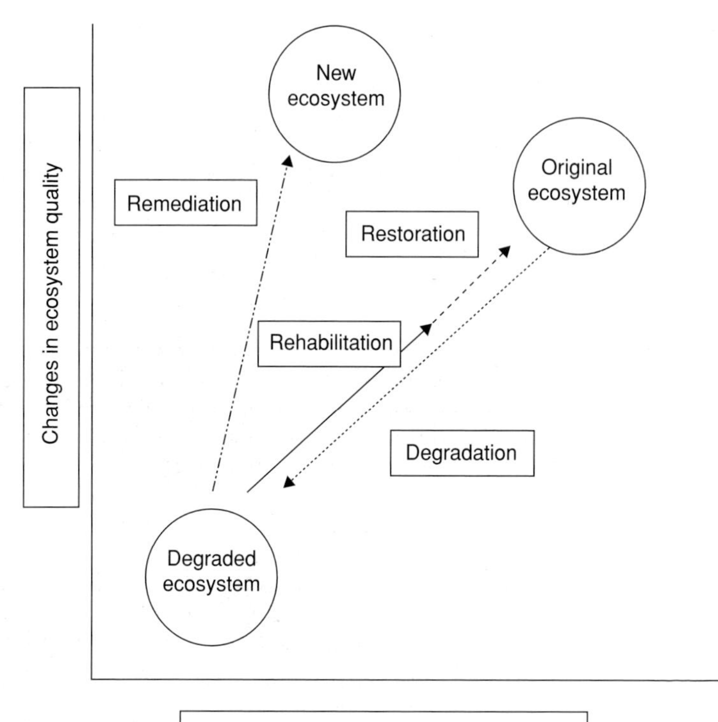

Figure 51 Restoration, rehabilitation and remediation

CASE STUDIES: THE COLE, THE SKERNE AND THE BREDE

The River Cole is one of three river restoration sites financially supported by LIFE, an EU fund which provides grant aid for schemes of environmental benefit (Figure 52). The others are the River Skerne in Darlington and the River Brede in Denmark. The aim of the RRP for the River Cole near Swindon is to change the water course, improve water quality and manage the bank-side vegetation. The project is being run by the RRP, the Environment Agency (formerly the National Rivers Authority), English Nature, the National Trust, the Countryside Commission and the EU.

The River Restoration Project was set up in 1994 with the main aim of establishing demonstration projects which showed how state-of-the-art restoration techniques could be used to re-create natural ecosystems in damaged river corridors. Three demonstration projects were set up, funded by European Union

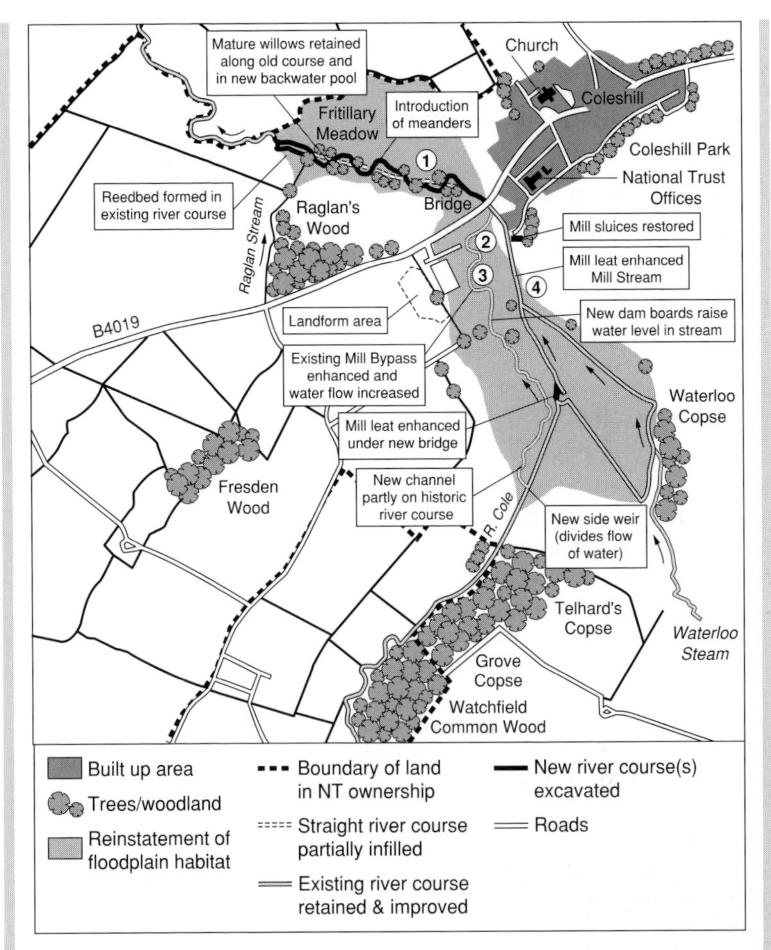

Figure 52 River Cole restoration site

LIFE money. The rural River Cole and the urban River Skerne have been restored over a 2 km reach, and were completed in 1996. Approximately 20 km of the rural River Brede was restored in Denmark.

There is a need to restore rivers because many have been seriously affected by urban and agricultural flood defences, land drainage and floodplain urbanisation. The result has been:

- extensive straightening and deepening of river channels, which has damaged wildlife habitats, reduced the value of fisheries and reduced much of the natural appeal of river landscapes
- major loss of floodplains and wetlands to intensive agriculture and urbanisation, which has destroyed floodplain habitats and

reduced the ability of floodplains to provide economically valuable functions, such as water and sediment storage
- rivers are used intensively as transport routes, carriers of waste disposal, for industrial purposes, water abstraction, recreation, etc.

There are two main ways of restoring rivers, natural and artificial. The first can take hundreds of years, consequently artificial restoration needs to take place. The benefits are greatest when natural channel forms, flows, sediment loads and floodplains are reinstated. However, structures that only copy natural features, such as a weir, give fewer benefits than the natural features.

Improving the River Cole

- Stretch 1. The river bed below Coleshill Bridge was raised to bring it back in line with its floodplain and to make it an important feature in the local landscape. This involved the introduction of more gravel riffles (fast flowing midstream ridges) and some small weirs (Figure 53).
- Stretch 2. The new river bed runs at the higher level at this length to fit in with the mill channel just upstream of the bridge. Rather than filling in part of the straightened river, a new meandering course was cut. Parts of the old course have been retained as backwaters to provide shelter for fish, birds and insects during high flows. This also means that neighbouring fields flood more frequently and help to recreate a water meadow.

Figure 53 River Cole meander

- Stretch 3. The restoration of the ancient course of the Cole appears to be possible at this site. Flood waters have restored the flood meadows along the western side of the mill.
- Stretch 4. The RRP hopes to restore the Cole Mill for occasional operation. The water levels in the Mill stream need to be raised for this to happen. The feeder stream (known locally as the Leat) is to be developed as a long lake with wet pasture and reed beds along its side.

Reed, willow and alder tree beds are very useful in cleansing streams which have been polluted by silt, fertiliser and treated sewage. A few carefully located beds of these plants are very effective at removing unwanted debris and pollutants.

Management

The overall aim of the proposals is to increase the extent to which the river and its floodplain interact, and to sustain a landscape that is rich in riverine and wetland wildlife. The key to success is the management of the floodplain, worked out in conjunction with local land managers. The main road at Coleshill Bridge and nearby buildings and sports fields have been protected from the increased risk of erosion and flooding.

Rivers and their floodplains are complex physical systems. However, they also have economic, social and political consequences. Balancing the physical demands of rivers with the economic and political demands that are based on the human use of the floodplain is difficult. The impact of human activity on natural systems is often very negative and it is impossible to imagine rivers and floodplains without wide-scale human activity. We are witnessing, perhaps, the dawn of a new era, one in which human activity is trying to restore rivers to their natural state rather than continually try to use and abuse them.

Benefits of restoration

- Nature conservation: wetland wildlife in the river and on the floodplain.
- Fisheries: species diversity and numbers.
- Water quality: increased interception of pollutants by vegetation and natural settling of sediments on the floodplain and river bed.
- Flood defence: additional flood storage is offered by the enlarged floodplain.
- Recreation: there is a strong public perception in favour of natural landscapes.

The River Skerne, Darlington

The River Skerne shares many characteristics with other urban streams:

- it has a high sediment load, especially of silt
- it is slow moving
- banks are overgrown with weeds
- many polluting developments such as factories and sewage works discharge into the river
- the floodplain contains a large amount of housing, roads, railways, factories and other industries.

The aim of the Skerne restoration project is to improve the quality of the river without reducing its function for flood defence (Figure 54). In particular, the Skerne will be improved by:

- the creation of new meanders in the river
- the introduction of sloping banks rather than vertical banks
- the growth of wetland plant species on the inside of meander belts
- strengthening the banks by planting trees and reeds
- creation of new wetland ecosystems
- improving the water quality from the sewage works
- creating a new footbridge so that access to the site is improved
- planting native species of plants to attract a richer, more diverse insect population.

River Brede

The River Brede flows through farmland in the low-lying county of South Jutland. It differs from the Cole in that the floodplain soils are much lighter sands and peats. Meanders had been removed from the river to create a straight course to enable intensive grassland farming. Weirs in the river, as well as the straightening, virtually eliminated a once valuable sea trout fishery.

A 5 km reach was re-meandered under the EU-LIFE project, but over 20 km of the Brede has now been restored as part of a nationwide strategy to improve the environmental management of river valleys. The scale of re-meandering is much greater than in the UK; the Brede once again meanders along the 500 m long floodplain and seasonal flooding has been restored to the valley.

The natural regeneration of the meandering river has been rapid and the sea trout are taking full advantage. As with the two UK sites, the progress of natural recolonisation is being closely monitored.

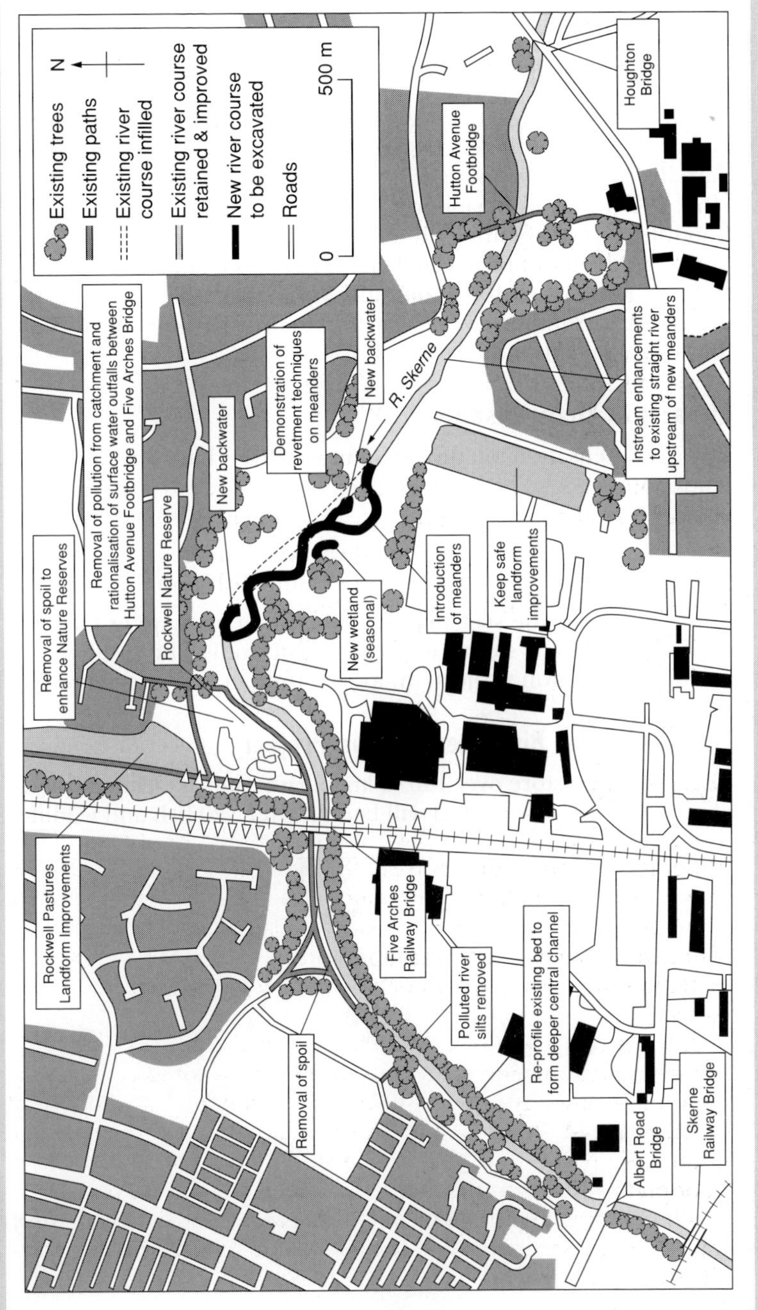

Figure 54 The River Skerne

CASE STUDY: CHANGING RIVER MANAGEMENT: THE KISSIMMEE RIVER

The Kissimmee (Figure 55) is a river in Florida that was adversely affected by hard engineering. To counter this, the river has been partially restored to some of its natural state. Between 1962 and 1971 the 165 km meandering Kissimmee River and flanking floodplain were channelised and thereby transformed into a 90 km, 10 m deep drainage canal. The river was channelised to provide an outlet canal for draining flood waters from the developing upper Kissimmee lakes basin, and to provide flood protection for land adjacent to the river.

Impacts of channelisation

The channelisation of the Kissimmee River had several unintended impacts:

- the loss of 30,000–35,000 acres of wetlands
- a reduction in wading bird and waterfowl usage
- a continuing long-term decline in game fish populations.

Concerns about the sustainability of existing ecosystems led to a state- and federally supported restoration study. The result was a massive restoration project, on a scale unmatched elsewhere.

The Kissimmee River Restoration Project

The aim is to restore over 100 sq km of river and associated floodplain wetlands. The project will benefit over 320 fish and wildlife

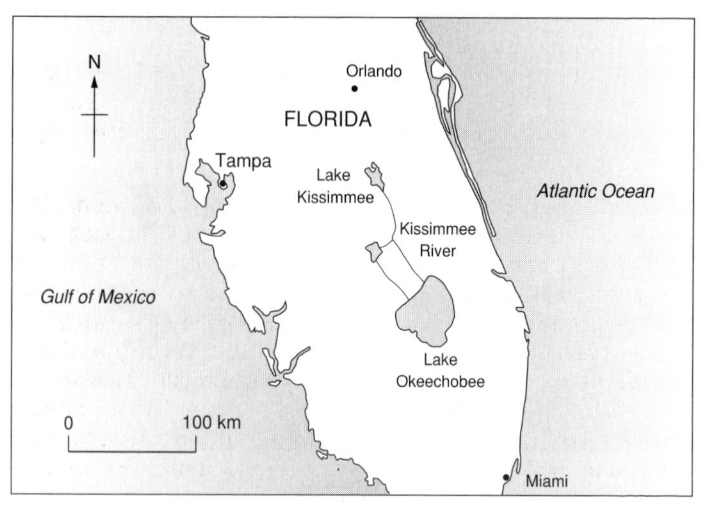

Figure 55 The Kissimmee River

species, including the endangered bald eagle, wood stork and snail kite. It will create over 11,000 ha of wetlands.

Restoration of the river and its associated natural resources requires dechannelisation. This entails backfilling approximately half of the flood control channel and re-establishing the flow of water through the natural river channel. In residential areas the flood control channel will remain in place.

The costs of restoration

It is estimated the project will cost $414 million (initial channelisation cost $20 million), a bill being shared by the state of Florida and the federal government. However, restoration, which began in 1999, will not be completed until 2010. Restoration of the river's floodplain could result in higher losses of water due to EVT during wet periods. Navigation may be impeded in some sections of the restored river; in extremely dry spells. It is, however, expected that navigable depths will be maintained at least 90% of the time.

Benefits of restoration

- Higher water levels should ultimately support a natural river ecosystem again.
- Re-establishment of floodplain wetlands and the associated nutrient filtration function is expected to result in decreased nutrient loads to Lake Okeechobee.
- It is possible that restoration of the Kissimmee River floodplain could benefit populations of key avian species, such as wading birds and waterfowl, by providing increased feeding and breeding habitats.
- Potential revenue associated with increased recreational usage (such as hunting and fishing) and ecotourism on the restored river could significantly enhance local and regional economies.

The world's largest wetland restoration project will spend US$700 million over two decades to revive the Florida Everglades It will include a series of six artificial wetlands known as 'storm water treatment areas', which will receive and clean up excess nutrients that enter the wetland from neighbouring farming districts'

Elsewhere, there has been mixed fortunes regarding river restoration in Australia, despite popular enthusiasm for restoration and rehabilitation. This is due, in part, to the lack of adequate planning and poorly defined objectives and strategies. Success of restoration projects appears to be influenced more by social and political factors rather than technical ones.

Summary

This chapter has looked at the management of rivers from a catchment perspective. We have also looked at methods of soft engineering, although a mix of hard and soft engineering is apparent in many rivers.

- The riparian zone is the land at the edge of the river.
- Management of the riparian zone is important for a number of reasons – streambank stability, ecology and water quality.
- Inland rivers are particularly difficult to manage.
- Conditions across many parts of Australia are deteriorating as years of unsuitable farming, deforestation, over-irrigation and drought begin to take effect.
- Catchment Management Plans are plans to manage rivers within their wider drainage basins.
- Agenda 21 refers to aspects of sustainable development that can be achieved at a local level.
- Even in urban streams there is potential for soft engineering.
- River restoration refers to the return of the river to its original natural state.
- Rehabilitation refers to a partial return to a river's original state.
- Remediation refers to small-scale improvements in a river's state.
- Restoration schemes may be small or large scale, urban or rural.
- Some rivers which have been altered by hard engineering have been partially restored – the Kissimmee is a good example.

Questions

1. Explain, with the use of examples, what is meant by the term 'soft engineering'.
2. Outlines the aims and methods of river restoration. For a restoration scheme that you have studied, comment on the effectiveness of the scheme.
3. Outline the importance of the riparian zone in water management.
4. Why are (**a**) urban rivers and (**b**) inland rivers difficult to manage?
5. To what extent can river management be sustainable?

6 Access to Water

KEY DEFINITIONS

Greenwater The rainfall that is stored in the soil and evaporates from it – is the main source of water for natural ecosystems, and for rainfed agriculture, which produces 60% of the world's food.
Blue water Renewable surface water runoff and groundwater recharge – is the main source for human withdrawals and the traditional focus of water resource management.

1 Introduction

In the past 200 years the world's population has grown exponentially – more people to be fed, and more water needed by each person for economic development. In the past 100 years the world population tripled, but water use for human purposes multiplied sixfold! Today perhaps half of all available freshwater is being used for human ends – twice what it was only 25 years ago.

The most obvious uses of water for people are drinking, cooking, bathing, cleaning and – for some – watering family food plots. This domestic water use, although crucial, is only a small part of the total. Worldwide, industry uses about twice as much water as households, mostly for cooling in the production of electricity. Far more water is needed for farming.

This has caused a water crisis. Sufficient clean water is essential to everyone's well-being. Over a billion people, nearly 20% of the world's population, do not have an access to safe and affordable drinking water, and 2.5 billion people, nearly 40% of the world's population, live in conditions lacking adequate sanitation. The vast majority of these people are in developing countries and the United Nations has identified water use as a priority for international aid. Insufficient supplies of water and sanitation disproportionately affect women, children and the poor. Most deaths from water-related diarrhoea are children under 15.

Access to water is now recognised as a key issue in development. Each year about 4 million people die of waterborne diseases, including 2 million children who die of diarrhoea. More than 800 million people, 15% of the world's population, is malnourished due in part to insufficient water for crops. The proportion of the world's population living in countries with significant water-stress will increase from approximately 34% in 1994 to 63% in 2025, including large areas of Africa, Asia and Latin America. More than 2.7 billion people will face severe shortages of freshwater by 2025 if the world keeps using water at today's rates.

Since the 1960s widespread acute water shortages have attracted increasing attention. For example, 1981–90 was declared the International Drinking Water Supply and Sanitation Decade. Following this, in 1992 the Earth Summit in Rio de Janeiro set goals for sustainable development, including guaranteeing every individual access to clean water and sanitation.

2 Main Uses of Water for Human Purposes

Water supply depends on several factors in the water cycle, including the rates of rainfall, evaporation, the use of water by plants (transpiration), and river and groundwater flows. It is estimated that less than 1% of all freshwater is available for people to use (the remainder is locked up in ice sheets and glaciers). Globally, around 12,500 cubic kilometres (km^3) of water are considered available for human use on an annual basis. This amounts to about 6600 m^3 per person per year.

Currently, the quantity of water used for all purposes exceeds 3700 km^3 per year. Agriculture is the largest user, consuming almost two-thirds of all water drawn from rivers, lakes and groundwater. While irrigation has undoubtedly contributed significantly to world agricultural production, it is extremely water intensive. Since 1960, water use for crop irrigation has risen by 60–70%. Industry uses about 20% of available water, and the municipal sector uses about 10%. Population growth, urbanisation and industrialisation have increased the use of water in these sectors.

The world's available freshwater supply is not distributed evenly around the globe, either seasonally, or from year to year. About three-quarters of annual rainfall occurs in areas containing less than one-third of the world's population, whereas two-thirds of the world's population live in the areas receiving only one-quarter of the world's annual rainfall. For instance, about 20% of the global average runoff each year is accounted for by the Amazon Basin, a vast region with fewer than 10 million people. Similarly, the Congo Basin accounts for about 30% of the Africa's annual runoff, but less than 10% of its population.

Throughout much of the developing world freshwater supply comes in the form of seasonal monsoon rains. These rains often run off too quickly for efficient use. India, for example, gets 90% of its rainfall during the summer monsoon season – at other times rainfall over much of the country is very low. Because of the seasonal nature of the water supply, many developing countries can use no more than 20% of their potentially available freshwater resources. Water supplies can also vary from year to year. Also, natural phenomena such as El Niño can lead to significant differences in rainfall in the southern Pacific Ocean, affecting South-east Asia and South and Central America.

Where water supplies are inadequate, two types of **water scarcity** affect LEDCs in particular:

- **physical water scarcity** where water consumption exceeds 60% of the usable supply. To help meet water needs some countries such as Saudi Arabia and Kuwait import much of their food and invest in desalinisation plants
- **economic water scarcity** occurs where a country physically has sufficient water resources to meet its needs, but additional storage and transport facilities are required – this will mean embarking on large and expensive water-development projects as in many in sub-Saharan countries.

In addition, in LEDCs, access to adequate water supplies is most affected by the exhaustion of traditional sources, such as wells and seasonal rivers. Access may be worsened by cyclical shortages in times of drought, inefficient irrigation practices and lack of resources to invest to meet demand and to increase the efficiency of irrigation systems. The International Water Management Institute estimates that 26 countries, including 11 in Africa, can be described as water scarce, and that over 230 million people are affected.

As world population and industrial output have increased, the use of water has accelerated, and this is projected to continue (Figure 56). By 2025 global availability of freshwater may drop to an estimated 5100 m^3 per person per year, a decrease of 25% on the 2000 figure. Rapid urbanisation results in increasing numbers of people living in

Use	1900	1950	1995
Agriculture			
Withdrawal	500	1100	2500
Consumption	300	700	1750
Industry			
Withdrawal	40	200	750
Consumption	5	20	80
Municipalities			
Withdrawal	20	90	350
Consumption	5	115	50
Reservoirs (evaporation)	0	10	200
Totals			
Withdrawal	600	1400	3800
Consumption	300	750	2100

Note: All numbers are rounded and given in cubic kilometres.

Figure 56 Global water use in the 20th century

urban shanty towns where it is extremely difficult to provide an adequate supply of clean water or sanitation.

As well as the need for an adequate quantity of water for consumption, it also needs to be of an adequate quality. However, WHO estimates that around 4 million deaths each year can be attributed to water-related disease, particularly cholera, hepatitis, dengue fever, malaria and other parasitic diseases. The incidence and effects of these diseases is most pronounced in developing countries, where 66% of people have no toilets or latrines.

Withdrawals for irrigation are nearly 70% of the total withdrawn for human uses – 2500 km^3 of 3800 km^3. Withdrawals for industry are about 20%, and those for municipal use are about 10%. Although we are withdrawing only 10% of renewable water resources, and consuming only about 5%, there are still problems for human use. Water is unevenly distributed in space and in time – and we are degrading the quality of much more water than we withdraw and consume.

In many LEDCs farmers use, on average, twice as much water per hectare as in industrialised countries, yet their yields can be three times lower – a sixfold difference in the efficiency of irrigation. Only about one-third of all the water used for agriculture actually contributes to making crops grow – of the rest some is returned to the system and reused but much is polluted or unusable.

Irrigation was first used about 6000 years ago, and transformed both the way in which land was used and human society. Food surpluses were created and stored, and centralised management and distribution networks emerged. Irrigation allowed new civilisations to blossom, such as those in the Tigris–Euphrates and the Nile (see *Climate and Society* for a fuller discussion of how climate change allowed civilisations to develop, prosper and collapse). In the long term societies based on irrigation fail for a number of reasons. These include environmental instability; soil salinity; declining soil fertility; declining crop productivity; dereliction of canals, levees and dams; the spread of disease; and competition for scarce water and land resources leading to regional warfare.

Unless carefully managed, irrigated areas risk becoming waterlogged and building up salt concentrations that could eventually make the soil infertile. Perhaps the biggest revolution in water resource management has been the small, cheap diesel or electric pump that gives farmers the means to invest in self-managed groundwater irrigation.

The high per capita residential use rates in North America and Europe have declined somewhat in recent years, in response to higher prices and environmental awareness. But in many sub-Saharan countries the average per capita use rates are undesirably low (10–20 litres/person/day) and need to be increased. A large share of the water withdrawn by households, services, and industry – up to 90% in areas where total use is high – is returned as wastewater, often in such a degraded state that major clean ups are required before it can be reused. There are three general ways to deal with water pollution:

- reduce releases from pollution sources
- transport pollutants to a place where they will do no harm
- treat the water to remove or convert the pollutants to harmless forms.

There is a wide range of technological options available to treat polluted water. These range from simple sand filters, through to sophisticated water treatment works that can remove a large number of potential contaminants, using technologies such as reverse osmosis and ion exchange. Reduction at source is the more environmentally preferable way of dealing with pollutants. In some locations natural ecosystems, such as wetlands and soils, are used as part of the treatment process.

Water pollution can also occur from the release of harmful substances, the use of agricultural chemicals, urban drainage, industrial effluents and saline intrusion as a result of over-pumping groundwater resources. A typical small treatment works can cost around £2–3 million annually. Currently, about 90% of wastewater in LEDCs is discharged without having undergone any treatment.

The real problem of drinking water and sanitation in developing countries is that too many people lack access to safe and affordable water supplies and sanitation. The WHO's World Health Report 1999 estimates that water-related disease caused 3.4 million deaths in 1998, more than half of them children.

Figure 57 summarises the numbers and proportions of the people in LEDCs lacking access to water and adequate sanitation in the year 2000. Figure 58 shows 'improved' technologies.

		World population (billions)		Proportion of world population	
	Total (billions)	with access (billions)	without access (billions)	with access (%)	without access (%)
Urban water supply	2.8	2.7	0.2	94	6
Rural water supply	3.2	2.3	0.9	71	29
Total water supply	6.1	5.0	1.1	82	82
Urban sanitation	2.8	2.4	0.2	86	14
Rural sanitation	3.2	1.2	2.0	38	62
Total sanitation	6.1	3.6	2.4	60	40

NB Figures have been rounded to the nearest 100 million

Figure 57 Global water supply and sanitation coverage

The following technologies are considered to be 'improved'	
Water supply	**Sanitation**
Household connection	Connection to a public sewer
Public standpipe	Connection to a septic system
Borehole	Pour-flush latrine
Protecting dug well	Simple pit latrine
Protecting spring	Ventilated improved pit latrine
Rainwater collection	

The following technologies were considered 'not improved'	
Water supply	**Sanitation**
Unprotected well	Service or bucket latrines (where excreta are manually removed)
Unprotected spring	Public latrines
Vendor-provided water	Open latrine
Bottled water	
Tanker truck provision of water	

Figure 58 Water supply and sanitation technologies considered to be 'improved' and 'not improved'

Figures 59 and 60 give the numbers and proportions with and without adequate access by region. Urban areas are better served than rural areas, and countries in Asia, Latin America and the Caribbean are better off than African countries. In the case of Asia, China and India, these alone comprised some 2280 million people in the year 2000, or over 60% of the region's total population. Many piped water systems, however, do not meet water quality criteria, leading more people to rely on bottled water brought in markets for personal use (as in major cities in Colombia, India, Mexico, Thailand, Venezuela and Yemen). Consumption of bottled water in Mexico is estimated at more than 15 billion litres a year, almost doubling between 1992 and 1998, and growing by 35% in 1996 and 1997 alone. Most of the 1.3 billion people living in poverty are women and children, the largest groups systematically under-represented in water resource management.

In some cases, the poor actually pay more for their water than the rich. For example, in Port-au-Prince, Haiti, surveys have shown that households connected to the water system typically paid around $1.00/m^3, while unconnected customers forced to purchase water from mobile vendors paid from $5.50 to a staggering $16.50/m^3.

Urban residents in the United States typically pay $0.40–0.80/m^3 for municipal water of excellent quality. In Lima, Peru, poor families on the edge of the city pay vendors roughly $3.00/m^3, 20 times the price for families connected to the city system.

	Total (billion)	Population		Proportion	
		With access (million)	Without access (million)	With access (%)	Without access (%)
Rural water					
Africa	490	230	260	47	53
Asia	2330	1730	600	75	25
LAC	130	80	50	62	38
Urban water					
Africa	300	250	50	85	15
Asia	1350	1250	100	93	7
LAC	390	360	30	93	7
Total population					
Africa	780	480	300	62	38
Asia	3680	2990	690	81	19
LAC	520	440	80	85	15

NB Figures have been rounded to nearest 100 million

Figure 59 Coverage of water supply by region

	Total (billion)	Population		Proportion	
		With access (million)	Without access (million)	With access (%)	Without access (%)
Rural sanitation					
Africa	490	220	270	45	55
Asia	2330	710	1620	31	69
LAC	130	60	70	49	51
Urban sanitation					
Africa	300	250	50	84	16
Asia	1350	1050	300	78	22
LAC	390	340	50	87	13
Total sanitation					
Africa	780	470	310	60	40
Asia	3680	1770	1910	48	52
LAC	520	400	120	78	22

NB Figures have been rounded to nearest 100 million

Figure 60 Access to rural sanitation by region

Residents of Jakarta, Indonesia, purchase water for \$0.09–0.50/m^3 from the municipal water company, \$1.80 from tanker trucks and \$1.50–2.50 from private vendors – as much as 50 times more than residents connected to the city system. Jakarta's water supply and disposal systems were designed for 500,000 people, but in 1985 the city had a population of nearly 8 million – and today, more than 15 million.

The city suffers continuous water shortages, and less than 25% of the population has direct access to water supply systems. The water level in what was previously an artesian aquifer is now generally below sea level – in some places 30 metres below. Saltwater intrusion and pollution have largely ruined this as a source of drinking water.

The picture is different for sanitation. Fewer people have adequate sanitation than safe water, and the global provision of sanitation is not keeping up with population growth. Between 1990 and 2000 the number of people without adequate sanitation rose from 2.6 billion to 3.3 billion. Least access to sanitation occurs in Asia (48%), especially in rural areas (31%). Nevertheless, there has been enormous progress since the 1970s. Over the 1980s, during the years of the Decade for Drinking Water and Sanitation, access to safe water more than doubled, while access to improved sanitation nearly tripled. Many of the improvements were relatively straightforward (Figure 58)

Progress in the 1990s was quantitatively not so large but was still impressive. For example, because of rapid increases in population, the proportions of people with access to water and sanitation grew more slowly. Urban areas are better served than rural areas. Nonetheless, by the year 2000, probably five or six times more people had access to safe water than 30 years before, and four or five times more people had access to adequate sanitation. China and India have each expanded coverage over the 1990s to reach about an additional 10% of their population.

There are still pressure points, especially in areas of rapid population growth. In many LEDCs rivers are now little more than open sewers. In Latin America only 2% of sewage receives any treatment at all. WHO says the numbers of people without sanitation will double to almost 5 billion within 23 years as the world becomes more urbanised. Contaminated water, inadequate sanitation and poor hygiene cause over 80% of all disease in developing countries. Half of all hospital beds in developing countries are full of people suffering waterborne diseases. Human waste is responsible for cholera, typhoid, trachoma, schistosomiasis and other infectious diseases that affect billions of people.

With squatter settlements in many of the world's poorest cities expanding rapidly, and local authorities unable to or legally prevented from providing sanitation, the situation is likely to deteriorate rapidly. Some 160,000 people are moving to cities from the countryside every day. At least 600 million people in Africa, Asia and Latin America now live in squatter settlements without any sanitation whatever, and governments are unable to cope. Meanwhile a UN Habitat report has warned that the lack of water in urban areas of Africa, the Middle East and Central Asia could lead to future conflicts. With a growing number of countries in Africa, the Middle East and Central Asia expected to face water scarcity, water could become a catalyst for regional conflicts, as oil did in the 1970s.

Figure 62 shows the projections of population growth and their implication for halving the proportion of people without access to water and adequate sanitation by 2015. The figures are large but they are achievable except in the areas of urban sanitation and rural water supply. The major challenges are those in Asia and Africa, not the Middle East nor Latin America.

Achieving the goals for water, sanitation and health is neither extremely difficult nor extremely costly. Examples exist that show how accelerated progress is possible and affordable. South Africa embarked on a major programme of safe drinking water in 1994. Within 7 years, by 2001, the numbers without safe water had been halved thus achieving the global goal 14 years ahead of 2015. The goal is now to achieve safe water for all by 2008.

Water and disease

Disease	Annual illness and deaths
Faecal-oral infections (waterborne and water-washed)	
Diarrhoea	1.5 billion cases for children under 5, 3.3 million deaths (5 million deaths, all ages)
Cholera	500,000 cases, 20,000 deaths
Typhoid fever	500,000 cases, 25,000 deaths
Ascariases (roundworm)	1.3 billion infected, 59 million clinical cases, 10,000 deaths
Water-washed infections (poor hygiene)	
Trachoma	146 million cases, 6 million people blind
Infection related to defective sanitation	
Hook worm	700 million infected

Source: Van der Hoek, Konradsen, and Jehangi 1999

Figure 61 Water-related diseases and death

CASE STUDY: DISEASE IN DHAKA

The best way into the Balurmath slum near the centre of Dhaka, Bangladesh, is over a rickety bamboo bridge that crosses a 7 metre ditch clogged deep with faeces, plastic bags and fetid liquids. The smell is evil, flies and mosquitoes swarm, and chickens root around in the excrement. Five thousand people live in Balurmath on a few acres of close-knit degradation, with no water supply, health services or anything beyond a few open latrines which empty directly into the ditch.

Most pay steep rents to an unofficial landlord, but live in fear that they will be forcibly evicted by the government, which owns the land and would like to develop it. According to one resident 'we all have coughs, fevers and diarrhoeal diseases. It has been like this for 10 years. A few water pumps were installed but they no longer work because the water level has gone down. The sanitation here is terrible, the smells are very bad. It is a 20 minute walk to get water'. Another claimed she needed 2 hours a day to collect water. Her husband earns £1/day collecting and recycling waste paper.

Sewage pollution is now one of the biggest and most common causes of illnesses, estimated to affect the health of more than 120 million people at any one time. In Asia, the level of sewage in rivers is 50 times higher than the UN guidelines. Diarrhoea is the world's second most serious killer of children, the World Health Organisation says, but in 90% of cases it can be easily prevented or treated.

WaterAid Bangladesh believe that it would cost only about £30,000 to supply this whole slum with decent water and latrines. That money, they claim, could completely transform the lives of thousands of people in less than a year, but DFID, the Department for International Development, has decided to pull out, saying it would be better to concentrate on rural communities.

WaterAid, Tearfund and a few other charities are working through their Bangladeshi partners to install simple water and sanitation in Dhaka's slums, but it is a herculean, underfunded and unfashionable task. More than 2.4 million people live in the city's slums, rising by more than 100,000 a year as Dhaka continues gain more and more informal settlements

But Balurmath is by no means the worst slum. Three miles away Outfall is one of the biggest, with 15,000 people living mostly in 2 × 3 metre rooms in very poor sanitary conditions. Yet, it would cost less than £60 to clean up the whole area round the well, and raise the pump on to a concrete plinth. £10,000 would provide decent latrines for the whole community, which is prepared to do the work itself but has no chance of raising the cash.

But the worst slum may be found right on the outskirts of the city, almost beneath a giant causeway built to prevent flooding. Here, on one side of the embankment, thousands of people live on bamboo stilt houses right above a giant cess pit. To go to the toilet, they simply open a hole in their floor. When it rains, the filth enters the houses.

Others have formed settlements on the far side of the road, but for months of the year they, too, must move out to avoid the seasonal floods. In the meantime, people drink water from a festering pond that serves also as their latrine and wash house. Children play in the drains and women queue for water.

> 'Cleaning up Dhaka's slums needs relatively little money, but sanitation is a dirty word with many donors and is given little political priority. The fact is, however, people cannot escape the stranglehold of poverty when they live in such squalor,' says WaterAid UK.
>
> (Adapted from *The Guardian*, 23 March 2002)

3 Managing Water Supplies

Water supply depends on many factors such as rainfall, temperature (for evaporation), vegetation, presence of rivers, groundwater and soil water stores, and use (or misuse) by human activities. The demand for water depends on a variety of factors including standards of living, availability and cost of water, ownership of water, type of economy (agricultural, industrial, tourist, etc.) and political disputes .

Water can be 'mined' in many ways. The main ones are:

- extraction from rivers and lakes (for example by primitive forms of irrigation such as the shaduf and Archimedes screw)
- it can be trapped behind dams and banks
- it can be pumped from aquifers (water-bearing rocks)
- desalinisation (changing saltwater to freshwater).

These ways of obtaining water can be achieved using either high-technology or low-technology methods.

Water harvesting refers to making use of available water before it drains away or is evaporated. Efficient use or storage of water can be achieved in many ways, for example

- irrigation of individual plants rather than of whole fields
- covering expanses of water with plastic or chemicals to reduce evaporation
- storage of water underground in gravel-filled reservoirs (again to reduce evaporation losses)

New approaches to irrigation

There has been a slow down in the growth of irrigated land. During the 1970s the annual expansion of irrigation was about 2% each year. Since 1982 that has fallen to 1.3%, and the expansion was taking place on less favoured sites, as the best sites had already been brought into farming. The expansion of irrigation between 2000 and 2020 is likely to be in the region of 0.6% annually.

At the start of the 21st century, the need for new approaches to irrigated farming is clear. One of the clearest signs of the problems associated with irrigation is the depletion of underground aquifers.

Over the last 30 years, the number of groundwater wells has expanded at an exponential rate, and groundwater depletion has changed from a small-scale isolated phenomenon to covering wide areas of cropland. The problem is widespread and found in, for example, central and northern China, north-west and southern India, parts of Pakistan, much of the south-west of the USA, North Africa, and the Middle East.

In Haryana, one of the most productive arable areas in India, water tables are dropping by 0.6–0.7 m/year, while in the Punjab it is about 0.5 m/year. The lack of groundwater could result in reduced food production in future. It is also leading widening rich-poor gaps in rural society. As the water table drops, farmers are forced to dig deeper wells and buy larger, more powerful pumps to get the water to the surface. In parts of the Punjab and Haryana wealthy farmers have used deep, expensive tubewells that cost up to £2000. This is clearly beyond the means of most small farmers. Denied the same access to water, they are condemned to producing smaller harvests, and so their income relative to rich farmers decreases.

The problem is widespread and reported from Pakistan, Bangladesh and China, too. Across much of the north China plain, which accounts for about 40% of China's grain, the water table has been dropping by as much as 1.5 m/year. The problem is not confined to LEDCs. In the USA, for example, several aquifers have been over-pumped. The most serious case is the Ogallala aquifer, which underlies parts of eight US states. Prior to extraction the aquifer held the equivalent of 200 years of flow of the Colorado River. So far only about 9% of the aquifer has been used up, but most of this has been on the High Plains of Texas. As a result of the subterranean resource, US agriculture has expanded into dry regions. However, up to 40% of cropland on top of the Ogallala may have to be abandoned by 2020 due to over-abstraction of the aquifer.

Water wars by sector

As water becomes scarcer, competition for it is increasing. This is not just between countries, but between economic sectors too. Rapidly growing cities and industries are increasingly looking towards irrigated agriculture as a source of water. A cubic metre of water used in China's industries generates more jobs and about 70 times more economic value than the same quantity used in farming. As supply shifts water will be switched to where it is more highly valued, and where it can make most profit. But there are costs associated with this. If 50% of the projected rise in urban and industrial demand for water is met by transferring water out of agriculture, grain production could drop by 300 million tonnes, 16% of the world's current harvest.

In other places, such as Australia, Central Asia's Aral Sea Basin and parts of the USA, there are pressures to transfer water back from irrigation to the natural environment. The states in Australia's

River basin/countries	Population 1999	Projected 2025 population	Change (%)
Aral Sea Kazakhstan, Kyrgyzstan, Tajikstan, Turkmenistan, Uzbekistan	56 million	74 million	+32
Ganges Bangladesh, India Nepal	1137	1631	+43
Jordan Gaza, Israel, Jordan Lebanon, Syria, West Bank	34	58	+71
Nile Burundi, DR Congo, Egypt, Eritrea, Ethiopia, Kenya, Rwanda, Sudan, Tanzania, Uganda	307	512	+67
Tigris-Euphrates Iraq, Syria, Turkey	104	156	+50

Figure 62 Populations in areas of potential water conflict and projected population growth

Murray-Darling river basin, for example, agreed to allocate 25% of the river's natural flow to maintaining the stream's ecological health.

Competition is also increasing between countries, as populations continue to grow in some of the world's water-short regions (Figure 62). In five of the world's problem water regions population growth by 2025 could reach a staggering 71%. Without water-sharing agreements, regional instability or war is a real possibility.

Water-stress is defined when a country's renewable water supplied drop below 1700 m^3 per capita. At present there are 34 countries in Africa, Asia and the Middle East that are classified as water-stressed, and all but two of them, Syria and South Africa, are net-importers of grain. These countries import about 50 million tonnes of grain each year, about one-quarter of the total traded internationally.

Quality rather than quantity

For most of the 20th century farmers have used irrigation water to raise land productivity (Figure 63). Essentially this means getting more crops from the land. However, the tide has turned somewhat and the focus of attention is now to get more benefit from every litre of water used in irrigation. Over half of the water taken from aquifers and rivers for irrigation never benefits a single plant but is lost through evaporation and seepage.

	Irrigated area (million hectares)	Share of cropland that is irrigated (%)
India	50.1	29
China	49.8	52
USA	21.4	11
Pakistan	17.2	80
Iran	7.3	39
Mexico	6.1	22
Russia	5.4	4
Thailand	5.0	24
Indonesia	4.6	15
Turkey	4.2	15
Uzbekistan	4.0	89
Spain	3.5	17
Iraq	3.5	61
Egypt	3.3	100
Bangladesh	3.2	37
Brazil	3.2	37
Romania	3.1	31
Afghanistan	2.8	35
Italy	2.7	25
Japan	2.7	62
World	255.5	17

(Source UN FAO Production yearbook, Rome)

Figure 63 Irrigated land worldwide

There is a long list of measures that can increase agricultural water productivity (Figure 64). Drip irrigation is one of the most untapped potential for farmers. Drip irrigation is a system of plastic tubes installed at or below the surface that deliver water to individual plants. The water, which can be enhanced with fertiliser, is delivered to the roots of plants, so that there is very little lost to evaporation. Drip irrigation can achieve as high as 95% compared with 50–70% for conventional flood systems. In surveys across the USA, Spain, Jordan, Israel and India, drip irrigation has been shown to cut water use by between 30% and 70%, and to increase crop yields by 20–90%, even leading to a doubling of productivity. Nevertheless, drip irrigation accounts for only about 1% of all irrigated land worldwide (Figure 65).

Expanding irrigation to small farmers

The benefits of irrigation have not been shared equally. The cheapest way to use groundwater, for example, is to use a diesel pump on a tubewell. These generally cost about £250, putting them out of the reach of small farmers. Lack of access to affordable technology for

Category	Measures
Technical	• Land levelling to apply water more uniformly • Surge irrigation to improve water distribution • Efficient sprinklers to apply water more uniformly • Low energy precision application sprinklers to cut evaporation and wind drift losses • Furrow diking to promote soil infiltration and reduce runoff • Drip irrigation to cut evaporation and other water losses and to increase crop yields
Managerial	• Better irrigation scheduling • Improving canal operations for timely deliveries • Applying water when most crucial to a crop's yield • Water-conserving tillage and field preparation methods • Better maintenance of canals and equipment • Recycling drainage and tail water
Institutional	• Establishing water user organisations for better involvement of farmers and collection of fees • Reducing irrigation subsidies and/or introducing conservation-orientated pricing • Establishing legal framework for efficient and equitable water markets • Fostering rural infrastructure for private-sector dissemination of efficient technologies • Better training and extension officers
Agronomic	• Selecting crop varieties with high yields per litre of transpired water • Intercropping to maximise use of soil moisture • Better matching crops to climate conditions and the quality of water available • Sequencing crops to maximise output under conditions of soil and water salinity • Selecting drought-tolerant crops where water is scarce or unreliable • Breeding water-efficient crop varieties

Figure 64 Options for improving irrigation water productivity

small farmers has been a major constraint on the spread of irrigation throughout much of sub-Saharan Africa.

There is, however, an array of irrigation technologies that are low cost, affordable and acceptable to small farmers in LEDCs (Figure 66). An example of one that has been very successful is the treadle pump, a human-powered irrigation. It is very similar to bicycles found in MEDC gyms, in which someone pedals on a machine while grasping the handles. However, in this case the pedalling causes groundwater to be sucked up into a cylinder.

Irrigation method	Typical efficiency	Water application needed to add 100 mm to root zone (%)	Water savings over conventional furrow irrigation (%)
Conventional furrow	60	167	–
Furrow with surge valve	80	125	25
Low-pressure sprinkler	80	125	25
Low energy precision Application (LEPA) Sprinkler	90–95	105	37
Drip	90–95	105	37

Figure 65 Efficiencies of selected irrigation methods, Texas High Plains

Substantial areas of valley bottomlands are suited for irrigation with low-cost pumps. Flood recession farming involves the planting of crops after a flood recedes.

Water harvesting aims at capturing and channelling a greater share of rainfall into the soil, and conserving moisture in the root zone where crops can use it.

In many drainage basins, constructing dams across stream head-waters can trap large amounts of runoff, which can either be channelled directly to a field, stored for later use in a tank or small reservoir, or allowed to percolate through the soil to recharge the groundwater. Such check dams are cheap to construct (generally less than £250) and typically last for about 5 years. Some are built entirely of local debris, while others may be built of manufactured bricks.

Other changes

Large government subsidies – about £20 billion each year – keep water prices artificially low, discouraging farmers from investing in efficiency improvements.

In many areas raising water prices is not popular, but there are alternatives to full-cost pricing such as tiered pricing and 'water banks'. A farmer has 'deposited' water in the bank while other users rent the water at a higher price, thereby giving the farmer a profit. If the farmer needs more water (s)he can borrow some from the bank, at a cost.

Other measures are, perhaps, less likely to occur. For example, if people ate less meat and ate more vegetables (i.e. adopt a more herbivorous lifestyle) the global ecosystem could support more people. The typical North American diet uses twice as much water in its production (in the field) as does the less meat-intensive diet of Asia and parts of Europe. In addition, further reductions in population growth

Technology or method	General conditions where appropriate	Examples
Cultivating wetlands, delta lands, valley bottoms; flood-recession cropping; rising flood cropping	Seasonally waterlogged floodplains or wetlands	Niger and Senegal river alleys; fadama of Nigeria; vdambos of Zambia and Zimbabwe; other parts of sub-Saharan Africa
Treadle pump, rower pump; pedal pump; rope pump; swing basket; Archimedean Screw; shaduf or beam and bucket; hand pump	Very small (less than 0.5 hectares) farm plots underlain by shallow groundwater or near perennial streams or canals in dry areas or areas with a distinct dry season	Easter India; Bangladesh; parts of southeast Asia; valley bottoms, dambos, fadama, and other wetlands of sub-Saharan Africa
Persian wheel; bullocks and other animal-powered pumps; low-cost mechanical pumps	Similar to those above, but where the average size of farm plots is roughly 0.5–2.0 hectares	Those above, in addition to parts of North Africa and Near East
Various forms of low-cost micro-irrigation, including bucket kits; drip systems; pitcher irrigation as well as microsprinklers	Areas with perennial but scarce water supply; hilly, sloping or terraced farmlands; tail-ends of canal systems; can apply to farms of various sizes, depending on the micro-irrigation technique	Much of northwest, central and southern India; Nepal; Central Asia, China, Near East; dry parts of sub-Saharan Africa; dry parts of Latin America
Tanks; check dams; percolation ponds; terracing; bunding; mulching; other water-harvesting techniques	Semi-arid and/or drought-prone areas with no perennial water source	Much of semiarid South Asia, including parts of India, Pakistan and Sri Lanka; much of sub-Saharan Africa; parts of China

Figure 66 Low-cost irrigation methods for small farmers

might ease pressures on water resources. On the other hand, critics argue that with reduced population growth there is an increase in standards of living, which are associated with an increased demand for many products, which increases the use of water.

Developers and water engineers are agreed that there are seven key areas of policy if water and hygiene development are to succeed:

- few LEDCs will be able to achieve the goals for hygiene, sanitation or water without widespread social and community action, which requires the mobilisation of populations

- involvement of the private sector is vital
- more use of simple, low-cost technologies and approaches which are affordable and acceptable are required
- better monitoring is needed with the results more publicly disseminated
- it is important to set a price policy which is politically acceptable and equitable
- it is vital to ensure the full participation of women in the management and operation of water, sanitation and hygiene programmes
- development aid must be more directly focused on water and sanitation for the poorest.

Summary

- Water use is increasing at an accelerating rate.
- Clean water is vital for human well-being.
- Lack of access to clean water is directly related to an increased risk of disease.
- There is a mismatch between the supply of water and the demand for it.
- Water supply per person is decreasing.
- Water scarcity disproportionately affects LEDCs.
- Farming is the main user of water.
- Access to water and sanitation vary between rural and urban areas. More people have access to water than sanitation – urban areas are better off than rural ones.
- In LEDC cities in particular, lack of access to clean water and sanitation is directly related to high rates of disease.
- There are a number of ways of improving the sustainable use of water – the greatest savings are likely to be in farming.
- The options chosen will depend on a variety of social, economic, political and environmental factors.
- Managing water resources has a local impacts as well as a global ones.

Questions

1. Describe and account for variations in access to clean water and sanitation.
2. Explain how human usage of water is not sustainable.
3. Explain why a lack of clean water is a barrier to development.
4. In what ways can water be used more sustainably in agriculture.

Bibliography

Australian Natural Resources Atlas (2000) National Land & Water Resources Audit, Canberra.

Cosgrove, WJ & Rijsberman, FR (2001) *Water Vision. Making Water Everybody's Business.* World Water Council.

Department for International Development (2001) *Addressing the Water Crisis.*

Global Water Partnership (2000) *Towards Water Security: A Framework for Action.*

Finlayson, M & Moser, M (Eds) (1991) *Wetlands.* International Waterfowl and Wetlands Research Bureau, Oxford/New York.

IPCC – Intergovernmental Panel on Climate Change (1996) *Climate Change 1995 – Impacts, Adaptations and Mitigation of Climate Change: Scientific Technical Analysis. Contribution of Working Group II to the Second Assessment Report of the IPCC.* Cambridge University Press, Cambridge.

Kirby, C & White, WR (Eds) *Integrated River Basin Development,* pp. 35–44. John Wiley, Chichester.

Lovett, S & Price, P (Eds) (1999) *Riparian Land Management Technical Guidelines, Volume One: Principles of Sound Management.* Land & Water Australia, Canberra.

Maltby, E (1986) *Waterlogged Wealth.* Earthscan, London.

Nagle, G (1999) *Britain's Changing Environment,* Focus on Geography series, Nelson, Cheltenham.

Price, P & Lovett, S (Eds.) (1999) *Riparian Land Management Technical Guidelines, Volume Two: On-ground Management Tools and Techniques.* Land & Water Australia, Canberra.

Thomas, DSG and Shaw, PA (1991) *The Kalahari Environment.* Cambridge University Press, Cambridge and New York.

United Nation Commission on Sustainable Development, *Comprehensive Assessment of the Freshwater Resources of the World.*

Wade M and Lopez-Gunn, E (1999) Wetland conservation, in Pacione, M (Ed) *Applied Geography: Principles and Practise.* Routledge, London.

Williams, M (Ed) (1990) *Wetlands: A Threatened Landscape.* Blackwells, Oxford.

Useful websites on water management:

Water aid at http://www.wateraid.org.uk

International Rivers Network at http://www.irn.org

World water forum at http://www.worldwaterforum.net

Nile Basin Initiative and information about the ten Nile Countries at http://www.nilebasin.org

Finding Solutions to Water Disputes atc http://www.gefweb.org

Convention on Wetlands (Ramsar, Iran, 1971): http://www.ramsar.org

Wicken Fen website: http://www.wicken.org.uk

United Nations Framework Convention on Climate Change. 1992. http://unfccc.int

Index